WATER AND CEREALS
IN DRYLANDS

P. Koohafkan
Director, Land and Water Division,
FAO, Rome

B.A. Stewart
Director, Dryland Agriculture Institute,
West Texas A&M University,
United States of America

Published by
The Food and Agriculture Organization of the
United Nations and Earthscan

publishing for a sustainable future

London • Sterling, VA

First published by The Food and Agriculture Organization of the United Nations and Earthscan in 2008

Paperback ISBN: 978-92-5-1060520 (FAO)
Paperback ISBN: 978-1-84407-708-3 (Earthscan)
Hardback ISBN: 978-1-84407-709-0 (Earthscan)

Printed and bound in Malta by Gutenberg Press Ltd

Cover photographs:
© FAO/J. VAN ACKER/13128
© FAO/G. TORSELLO/17398
© FAO/G. BIZZARRI/19715

For a full list of publications please contact:

Earthscan
Dunstan House
14a St Cross Street, London EC1N 8XA, UK
Tel: +44 (0)20 7841 1930
Fax: +44 (0)20 7242 1474
Email: earthinfo@earthscan.co.uk
Web: www.earthscan.co.uk

22883 Quicksilver Drive, Sterling, VA 20166-2012, USA

Earthscan publishes in association with the International Institute for Environment and Development

A catalogue record for this book is available from the British Library

Library of Congress Cataloging-in-Publication Data been applied for

Mixed Sources
Product group from well-managed
forests, and other controlled sources
www.fsc.org Cert no. TT-CoC-002424
© 1996 Forest Stewardship Council

The paper used for this book is FSC-certified.
FSC (the Forest Stewardship Council) is an
international network to promote responsible
management of the world's forests.

CONTENTS

LIST OF TABLES

LIST OF FIGURES

LIST OF BOXES

LIST OF PLATES

LIST OF ACRONYMS

CA	Conservation agriculture
CBT	Conservation bench terrace
CRP	Conservation Reserve Program
GHG	Greenhouse gas
GIS	Geographical information system
HYV	High-yielding variety
IFPRI	International Food Policy Research Institute
LAI	Leaf area index
LGP	Length of growing period
PET	Potential evapotranspiration
PES	Payment for Environmental Services
SOM	Soil organic matter
SWC	Soil Water Conservation
UNCCD	United Nations Convention to Combat Desertification
WH	Water harvesting

ACKNOWLEDGEMENTS

This publication is the result of several years of research, studies and field work of the authors enriched by discussions, interactions and suggestions of numerous scientists and practitioners across the globe.

The Authors are particularly grateful to Dr. Robert Brinkman and Mrs. Anne Woodfine for the editing of the original manuscript and to Mrs. Karen Frenken, Mr. Jan Poulisse and Mrs. Ines Beernaerts for their comments and suggestions. Authors would also like to thank the Peer Review Committee Dr. John Ryan, Dr. Johan Rockstrom and Dr. Suhaj Wani for their suggestions and contributions. Finally, thanks go to, Ms Mary Jane de la Cruz, Mr. Simone Morini and Mrs. Bouchra El-Zein for their assistance for formatting and designing of the book.

The world's food supply is obtained either directly or indirectly from the abundance of plant species, but fewer than 100 are used for food. Worldwide, about 50 species are cultivated actively, and as few as 17 species provide 90 percent of human food supply and occupy about 75 percent of the total tilled land on earth. Eight cereal grains –wheat, barley, oat, rye, rice, maize, sorghum and millet– provide 56 percent of the food energy and 50 percent of the protein consumed on earth (Stoskopf, 1985). Cereals continue as the most important source of total food consumption in the developing countries where direct consumption of grains provides 54 percent of total calories and 50 percent for the world as a whole (FAO, 2006). Wheat and rice are by far the most widely consumed cereals in the world. Maize is a major crop for both direct and indirect human consumption because it is a major energy feed for animals. Wheat, rice, and maize make up approximately 85 percent of the world's production of cereals.

As food and water needs continue to rise, it is becoming increasingly difficult to supply more water to farmers. The supply of easily accessible freshwater resources is limited both locally and globally. In arid and semiarid regions, in densely populated countries and in most of the industrialized world, competition for water resources has set in. In major food-producing regions, scarcity of water is spreading due to climate change and increased climate variability. In light of demographic and economic projections, the freshwater resources not yet committed are a strategic asset for development, food security, the health of the aquatic environment and, in some cases, national security. In large parts of the drylands, no irrigation water is available – rainfed crop and pasture yields are both low and uncertain. Runoff, evaporation and deep percolation from the soil surface drastically reduce the proportion of rainfall available for plant growth. However, even small amounts of additional water would significantly increase yields in drylands at very high water-use efficiencies, if other factors – including plant nutrient availability – were adequate. Several approaches can make such additional water available to crops and pasture from the local rainfall with low-cost low-risk land and water management techniques. Runoff can be used more productively and infiltration increased in arid areas by pitting or tied ridges, and by increasing surface roughness. In semi-arid and dry subhumid areas, maintaining a cover of crops or crop residues on the soil throughout the year in a zero- or minimum-tillage system can be even more effective. Recent experiences of conservation agriculture bring about multiple benefits for farmers while addressing local and global environmental concerns (Pretty & Koohafkan, 2002). Where such measures still do not provide the crop with adequate moisture throughout the growing period, water-harvesting approaches such as "runoff farming" to supplement rainfall on a smaller area may be viable options.

Where rainfall is distributed sparsely throughout the year, dry farming may be an option. This approach entails capturing rainfall during a fallow period and storing it in the soil for use during the subsequent cropping period. Storage efficiency can be increased by reduced tillage or no-tillage where crop residues can be maintained on the soil surface as mulch.

This volume discusses the drylands and their land uses, with an emphasis on cereal production. It includes an outline on the recent development of competing use of cereals for the production of ethanol biofuel. This paradigm shift could have far-reaching consequences (social, environmental and for food security), potentially encouraging production in even more marginal lands. The volume touches on the roles of livestock, placing the various technologies and practises that enhance water availability to crops in drylands in their technical, agro-ecological and socio-economic perspective. The predicted future impacts of human-induced climate change on dryland systems are briefly noted.

I therefore appeal to the international community to join FAO in its continuing efforts towards alleviating poverty and hunger through the promotion of agricultural development, the improvement of nutrition and the pursuit of food security throughout the world. With your help, success is at the end of our efforts, perseverance and commitment.

Jacques Diouf
FAO DIRECTOR-GENERAL

World food crop production has more than kept pace with the rapid growth in population in the past four decades. World population increased from 3 billion to 6.6 billion between 1960 and 2006 (UNFPA, 2008), food consumption measured in kcals per capita increased from 2 358 to 2 803 between 1964 and 1999 (WHO, 2008), and food prices fell between 1961 and 1997 by 40 percent in real terms (FAOSTAT, 2007). However, these global statistics do not fully reflect the wide range of differences between and within individual countries.

Cereals are by far the most important source of total food consumption in the developing countries. Direct food consumption of cereals in these countries provides 54 percent of total calories and 50 percent for the world as a whole (FAO, 2006). There are, however, wide variations among countries. Only 15-30 percent of total calories are derived from cereals in countries where roots and tubers are dominant (e.g. Rwanda, Burundi, the two Congos, Uganda, Ghana, etc.) and in high-income countries with predominantly livestock-based diets (e.g. U.S., Canada, Australia, etc.). These latter countries, however, consume large amounts of cereals indirectly in the form of animal feed for the livestock products consumed as food. Approximately 37 percent of the world's cereals are used for feed (FAOSTAT, 2007). Production of cereals increased 2.6 times between 1961 and 2005 compared to an increase in population of 2.1 times. This increase in per capita cereal production alleviated hunger problems in several countries and contributed to an increase in meat consumption in developing countries from about 9 kg per capita in 1961 to more than 30 kg per capita in 2005 (FAOSTAT, 2007).

The growth rate of cereal production has slowed in recent years. Production of cereals grew at a rate of 3.7 percent per annum during the 1960s, but slowed to 2.5 percent, 1.4 percent, and 1.1 percent per annum during the subsequent three decades to 2001 (FAO, 2006). Per capita food use of cereals seems to have peaked in the early 1990s, and this is true for the world as a whole. World conumption per capita fell from 171 kg/person/year in 1989/91 to 165 kg in 1999/01; in the developing countries , from 174 kg to 166 kg (FAO, 2006). It is not clear why cereal consumption had decreased in developing countries when so many of them are far from having reached adequate levels of food consumption. The use of cereals for all uses reached 329 kg/person/year in 1989/91 and fell to 309 kg in 1999/01. FAO (2006) projects that total usage in 2050 will be 339 kg. Even though per capita cereal usage has declined slightly, and world population growth has slowed somewhat from earlier forecasts, FAO forecasts that cereal production will need to increase from 1.9 billion tonnes in 2001 to 3 billion tonnes by 2050. This is a challenge that should not be taken lightly in view of the increasing pressures and competition for soil and water resources.

Recently, the future of cereal production and consumption has changed dramatically since the rising cost of fossil fuels and the need for greener energy use has resulted in some cereals (principally maize – also sorghum and wheat) being used to produce ethanol for fuel. Although it is too early to determine the long-term impact that this development will have on the supply and cost of food, the fact that both food and fuel systems are competing for cereals is likely a profound development that could have unintended environmental, social and food security consequences of major importance. Although energy prices tend to influence the food and agriculture sector because of the effect on the price of fertilizer, fuel, transportation, etc., it is only recently that the price of grain has been directly linked to the price of oil. Brown (2008) states that in this new situation the world price of grain is drifting up toward its energy-equivalent value, since when the fuel value of a crop exceeds its food value, the crop will enter bio-fuel production.

Wheat, rice and maize make up about 85 percent of the world's cereal production (FAOSTAT, 2007). Wheat and rice are by far the most widely consumed cereals in the world in the direct form, while maize is important for both direct and indirect human consumption as a major feed ingredient for animals. Cereals are grown on 49 percent of the world's harvested area, with 65 percent being grown in developing countries in 2005 (FAO, 2006). World population, currently 6 700 million, is projected to reach 8 000 million by 2025, with more than 97 percent of the growth occurring in developing countries (FAO, 2003c). Cereals will continue to be an important food source, particularly in developing countries and consequently, it is vital that production continues at a pace to match consumption. If significant volumes of cereals produced in developed countries are used for ethanol production, the price of cereals is likely to increase and the amounts available for export and for reserve are likely to decrease. Brown (2008) estimated that one fifth of the entire U.S. grain harvest in 2007 will be used to produce ethanol. This will likely increase the necessity for developing countries to become more self-sufficient in cereals.

The agronomic technologies that have allowed steady increases in world food production to date have largely been based on high-yielding varieties (HYVs), fertilizers, pest control and irrigation. Irrigation has been particularly important in developing countries, where the total irrigated area increased from 102 million ha in 1961 to 208.7 million ha in 2002. This compares with 37 million ha in the developed countries in 1961 and 68 million ha in 2002 (Molden et al., 2007; Svendsen and Turral, 2007). Worldwide, 19.7 percent of arable land is irrigated, and it contributes 40 percent of total agricultural production (Svendsen and Turral, 2007). Irrigated agriculture is responsible for approximately 70 percent of all the freshwater withdrawn in the world (Molden and Oweis, 2007). In most developing countries, agriculture accounts for 80 percent of water use (UNDP, 2006). The amount varies widely among countries, however, ranging from more than 90 percent in agricultural economies in the arid and semi-arid tropics to less than 40 percent in industrial economies in the humid temperate regions (FAO, 1996a).

In 1996, 48 percent of cereal production in developing countries (excluding China) came from irrigated lands (FAO, 1996a). Ringler et al. (2003) reported that 38 percent of the cereals grown in developing countries in 1995 were on irrigated lands and accounted for 60 percent of cereal production. This is in contrast to many developed countries where cereals are largely grown without irrigation. An estimated 60 percent of the wheat produced in developing countries is irrigated, while only 7 percent of the wheat and 15 percent of the maize are irrigated in the United States of America (USDA, 1997). Brown (2008) estimated that one fifth of U.S grain harvest comes from irrigated land compared to three fifths for India and four fifths for China. The other major cereal-producing areas of developed countries (in Canada, Australia and Europe) are also predominantly non-irrigated systems.

The development of additional irrigated land in developing countries will need to continue in order for food production to keep pace with population growth and to ensure regional food security. This should be coupled with increased water productivity from improved irrigation and water management. However, there are also effective alternatives to irrigation, even in less humid areas, through the development of dryland farming systems.

With the cost of developing additional irrigated lands often ranging from US$2 000 to more than 17 000/ha (AQUASTAT, 2008), it is imperative that alternative water-management and production systems be considered for at least a part of the anticipated future food demand. Huge capital sums have already been invested to develop irrigation. In comparison, there has been very limited investment focused on raising the overall productivity of drylands. This is understandable because the benefits of irrigated agriculture are significant, immediate, predictable and dependable. In contrast, cereal yields in dryland regions where irrigation is not an option typically range from zero to three or more times the average yield. This high variability limits the effectiveness of inputs such as fertilizers and pesticides, resulting in a high economic risk associated with their use.

However, several soil- and water-management options, such as conservation agriculture, run-off farming and dry farming using fallow storage can increase soil moisture in dryland areas, increasing yields and reducing their variability. Some preliminary estimates show that the average yield of rainfed cereals in drylands could be increased by 30–60 percent by making available an additional 25–35 mm of water to crops during critical growth periods through water conservation and harvesting. These benefits are attainable in most dryland areas of the world and justify investment in water conservation and water harvesting. There are also social and environmental benefits which support the investment costs in water conservation and harvesting far beyond those strictly related to increased grain production.

The objectives of this study were to:

* Emphasize the importance of dryland development for future food production (particularly cereals), food security and poverty alleviation;
* Present water-conservation and water-harvesting approaches and investment options that can increase cereal production in dryland regions;
* Suggest policies for more efficient use of existing natural resources in order to lessen the dependence of agriculture on further irrigation development.

Chapter 1 characterizes and discusses drylands and their land uses, highlighting their importance to the growing populations who occupy them. Chapter 2 reviews global trends in cereal production and considers the constraints on the further expansion of irrigation. Chapter 3 reviews water-conserving technologies and practices for enhancing cereal production in drylands by more integrated and efficient use of existing land and water resources. Chapter 4 reviews some wider environmental issues relating to water harvesting and soil water conservation in drylands. Chapter 5 considers some of the social and economic benefits that result from investing in water-conservation and water-harvesting systems in dryland areas as well as investment constraints and potential. The study concludes with policy considerations and recommendations for future actions.

CHAPTER 1

Drylands, people and land use

CHARACTERISTICS OF DRYLANDS

There is no single agreed definition of the term drylands. Two of the most widely accepted definitions are those of FAO and the United Nations Convention to Combat Desertification (UNCCD, 2000). FAO has defined drylands as those areas with a length of growing period (LGP) of 1–179 days (FAO, 2000a); this includes regions classified climatically as arid (Plate 1), semi-arid and dry subhumid. The UNCCD classification employs a ratio of annual precipitation to potential evapotranspiration (P/PET). This value indicates the maximum quantity of water capable of being lost, as water vapour, in a given climate, by a continuous stretch of vegetation covering the whole ground and well supplied with water. Thus, it includes evaporation from the soil and transpiration from the vegetation from a specific region in a given time interval (WMO, 1990). Under the UNCCD classification, drylands are characterized by a P/PET of between 0.05 and 0.65.

According to both classifications, the hyperarid zones (LGP = 0 and P/PET < 0.05), or true deserts, are not included in the drylands and do not have potential for agricultural production, except where irrigation water is available.

While about 40 percent of the world's total land area is considered to be drylands (according to the UNCCD classification system), the extent of drylands in various regions ranges from about 20 percent to 90 (Table 1 and Figure 1).

Drylands are a vital part of the earth's human and physical environments. They encompass grasslands, agricultural lands, forests and urban areas. Dryland ecosystems play a major role in

global biophysical processes by reflecting and absorbing solar radiation and maintaining the balance of atmospheric constituents (Ffolliott *et al.*, 2002). They provide much of the world's grain and livestock, forming the habitat that supports many vegetable species, fruit trees and micro-organisms.

High variability in both rainfall amounts and intensities are characteristics of dryland regions, as are the occurrence of prolonged periods of drought. A drought is defined as a departure from the average or normal conditions, sufficiently prolonged (1-2 years - FAO, 2004) as to affect the hydrological balance and adversely affect ecosystem functioning and the resident populations. There are actually four different ways that drought can be defined (National Weather Service, 2004). Meteorological drought is a measure of the departure of precipitation from normal. Due to climatic differences, a drought in one location may not be a drought in another location. Agricultural drought refers to situations where the amount of soil water is no longer sufficient to meet the needs of a particular crop. Hydrological drought occurs when surface and subsurface water supplies are below normal. Socioeconomic drought describes the situation that occurs when physical water shortages begin to affect people. This report is primarily concerned with agricultural droughts.

The terms drought and aridity are sometimes used interchangeably, but they are different.

TABLE 1
Regional extent of drylands

REGION	ARIDITY ZONE							
	Arid		Semi-arid		Dry subhumid		All drylands	
	(1 000 km²)	(%)	(1 000 km²)	(%)	(1 000 km²)	(%)	(1 000 km²)	(%)
Asia (incl. Russia)	6 164	13	7 649	16	4 588	9	18 401	39
Africa	5 052	17	5 073	17	2 808	9	12 933	43
Oceania	3 488	39	3 532	39	996	11	8 016	89
North America	379	2	3 436	16	2 081	10	5 896	28
South America	401	2	2 980	17	2 223	13	5 614	32
Central America and Caribbean	421	18	696	30	242	10	1 359	58
Europe	5	0	373	7	961	17	1 359	24
World total	**15 910**	**12**	**23 739**	**18**	**13 909**	**10**	**53 558**	**40**

Source: UNSO/UNDP, 1997.

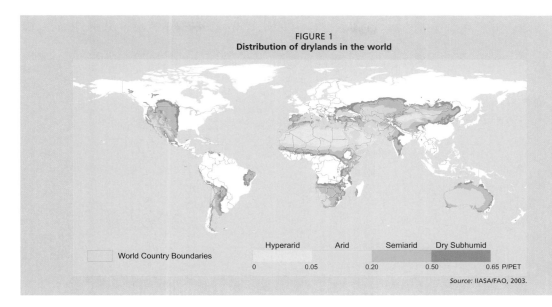

FIGURE 1
Distribution of drylands in the world

World Country Boundaries

Hyperarid   Arid   Semiarid   Dry Subhumid

0   0.05   0.20   0.50   0.65 P/PET

Source: IIASA/FAO, 2003.

Aridity refers to the average conditions of limited rainfall and water supplies, not to the departures from the norm, which define a drought. All the characteristics of dryland regions must be recognized in the planning and management of natural and agricultural resources (Jackson, 1989). Because the soils of dryland environments often cannot absorb all of the rain that falls in large storms, water is often lost as runoff (Brooks *et al.*, 1997). At other times, water from a rainfall of low intensity can be lost through evaporation when the rain falls on a dry soil surface. Molden and Oweis (2007) state that as much as 90 percent of the rainfall in arid environments evaporates back into the atmosphere leaving only 10 percent for productive transpiration. Ponce (1995) estimates that only 15 to 25 percent of the precipitation in semiarid regions is used for evapotranspiration and that a similar amount is lost as runoff. Evapotranspiration is the sum of transpiration and evaporation during the period a crop is grown. The remaining 50 to 70 percent is lost as evaporation during periods when beneficial crops are not growing.

Three major types of climate are found in drylands: Mediterranean, tropical and continental (although some places present departures from these). Dryland environments are frequently characterized by a relatively cool and dry season,

followed by a relatively hot and dry season, and finally, by a moderate and rainy season. There are often significant diurnal fluctuations in temperatures which restricts the growth of plants within these seasons.

The geomorphology of drylands is highly variable. Mountain massifs, plains, pediments, deeply incised ravines and drainage patterns display sharp changes in slope and topography, and a high degree of angularity. Streams and rivers traverse wide floodplains at lower elevations and, at times, are subject to changes of course, often displaying braided patterns. Many of these landforms are covered by unstable sand dunes and sand sheets. Dryland environments are typically windy, mainly because of the scarcity of vegetation or other obstacles that can reduce air movement. Dust storms are also frequent when little or no rain falls.

Soils in drylands are diverse in their origin, structure and physicochemical properties. In general, they include Calcisols, Gypsisols, Leptosols and Steppe soils (FAO, 2004) (Figure 2). Important features of dryland soils for agricultural production are their water holding capacity and their ability to supply nutrients to plants. As there is little deposition, accumulation or decomposition of organic material in dryland environments, the

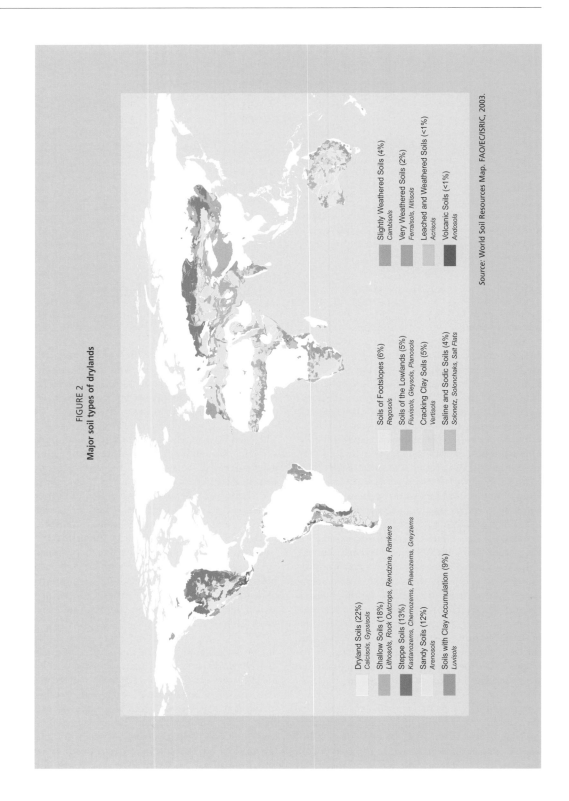

FIGURE 2
Major soil types of drylands

Dryland Soils (22%)
Calcisols, Gypsisols

Shallow Soils (18%)
Lithosols, Rock Outcrops, Rendzina, Rankers

Steppe Soils (13%)
Kastanozems, Chernozems, Phaeozems, Greyzems

Sandy Soils (12%)
Arenosols

Soils with Clay Accumulation (9%)
Luvisols

Soils of Footslopes (6%)
Regosols

Soils of the Lowlands (5%)
Fluvisols, Gleysols, Planosols

Cracking Clay Soils (5%)
Vertisols

Saline and Sodic Soils (4%)
Solonetz, Solonchaks, Salt Flats

Slightly Weathered Soils (4%)
Cambisols

Very Weathered Soils (2%)
Ferralsols, Nitisols

Leached and Weathered Soils (<1%)
Acrisols

Volcanic Soils (<1%)
Andosols

Source: World Soil Resources Map. FAO/EC/ISRIC, 2003.

organic content of the soils is low and, therefore, natural soil fertility is also low.

Much of the water that is available to people living in drylands regions is found in large rivers that originate in areas of higher elevation (e.g. the Nile, Tigris-Euphrates, Indus, Ganges, Senegal, Niger and Colorado Rivers). Groundwater resources can also be available to help support development. However, the relatively limited recharge of groundwater resources is dependent largely on the amount, intensity and duration of the rainfall and soil properties, the latter including the infiltration capacity and waterholding characteristics of the soil, which also influence the amount of surface runoff. With current management practices, much of the rainfall is lost by evapotranspiration or runoff. As a result, groundwater is recharged only locally by seepage through the soil profile. Surface runoff events, soil-moisture storage, and groundwater recharge in dryland regions are generally more variable and less reliable than in more humid regions. In some areas, important reservoirs of fossil ground water exist and continue to be used by human population. Fossil water is groundwater that has remained in an aquifer for millennia. Extraction of fossil groundwater is often called mining because it is a non-renewable resource. For such aquifers, including the vast U.S. Ogallala aquifer, the deep aquifer under the North China Plain, and the Saudi aquifer, depletion can bring pumping to an end. In some cases, farmers can convert to dryland farming, but in arid regions it is the end of farming. Brown (2008) cited a 2001 China study that showed the water table under the North China Plain is falling fast and this area produces over half of the country's wheat and a third of its maize. He also cited a World Bank study that reported 15 percent of India's food supply is produced by mining groundwater. Even in areas where groundwater is recharged, it is frequently used at rates that exceed the recharge rate. Water that is available for use in many drylands regions can be affected also by salinity and mineralization (Armitage, 1987).

DRYLANDS PEOPLE

Drylands are inhabited by more than 2 000 million people, nearly 40 percent of the world's population (White and Nackoney, 2003).

Dryland populations are frequently some of the poorest in the world, many subsisting on less than US$1 per day (White et al., 2002).

The population distribution patterns vary within each region and among the climate zones comprising drylands. Regionally, Asia has the largest percentage of population living in drylands: more than 1 400 million people, or 42 percent of the region's population. Africa has nearly the same percentage of people living in drylands (41 percent) although the total number is smaller at almost 270 million. South America has 30 percent of its population in drylands, or about 87 million people (Table 2).

Rural people living in drylands can be grouped into nomadic, semi-nomadic, transhumant and sedentary smallholder agricultural populations. Nomadic people are found in pastoral groups that depend on livestock for subsistence and, whenever possible, farming as a supplement. Following the irregular distribution of rainfall, they migrate in search of pasture and water for their animals. Semi-nomadic people are also found in pastoral groups that depend largely on livestock and practice agricultural cultivation at a base camp, where they return for varying periods. Transhumant populations combine farming and livestock production during favourable seasons, but seasonally they might migrate along regular routes using vegetation growth patterns of altitudinal changes when forage for grazing diminishes in the farming area. Sedentary (smallholder) farmers practise rainfed or irrigated agriculture (Ffolliott et al., 2002) often combined with livestock production.

The human populations of the drylands live in increasing insecurity due to land degradation and desertification and as the productive land per capita diminishes due to population pressure (Plate 2). The sustainable management of drylands is essential to achieving food security and the conservation of biomass and biodiversity of global significance (UNEP, 2000).

LAND USE SYSTEMS IN DRYLANDS

Dryland farming is generally defined as farming in regions where lack of soil moisture limits crop or pasture production to part of the year. Dryland farming systems are very diverse, including a variety of shifting agriculture systems, annual croplands, home gardens and mixed agriculture–livestock systems, also nomadic pastoral and transhumant systems (Figure 3 and Plate 3). They also include fallow systems and other indigenous intensification systems (FAO, 2004) for soil moisture and soil fertility restoration. Haas, Willis and Bond (1974) defined fallow as a farming practice where and when no crop is grown and all plant growth is controlled by tillage or herbicides during a season when a crop might normally be grown

The major farming systems of the drylands vary according to the agro-ecological conditions of these regions. A recent study of the Land Degradation Assessment in Dryland projects (LADA, 2008) identified the major farming systems in drylands according to socio-economic information, agro-ecology and possibilities for irrigation. The majority of the drylands used for agriculture is under cereal cultivation. Annex 2 summarizes farming practices in some of the major dryland areas of the world.

Successful dryland farming requires the integrated management of soil, water, crops and plant nutrients. Small-scale, resource-poor, usually subsistence-based farmers, widely referred to

as small-holders, operate and survive in these varied, changeable and hazardous environments by being able to manage the multiple risks (FAO, 2004) through diversification, flexibility and adaptability (Mortimore and Adams, 1999). Stewart and Koohafkan (2004) and Stewart, Koohafkan and Ramamoothy (2006) have also reviewed the importance and some of the constraints of dryland farming. Expansion of cropland areas in dryland regions can fail owing to overexpansion of inappropriate production technologies into the drylands environment. Increased population pressures and human expansion into drier areas during long wet periods leave an increasing number of people vulnerable to drought. Removing critical production elements (e.g. dry-season grazing areas) from the traditional complex land-use systems through the introduction of irrigated and non-irrigated crops, or the increased industrial and urban use of water, break links in traditional production chains.

One of the reasons why dryland farming has generally been inefficient is the poverty trap. Resource-poor people living in marginal environments try to survive by avoiding damage resulting from hazards. Avoiding risks often entails maximizing the use of labour while minimizing the use of capital-intensive resources as the poor cannot afford to invest sufficiently

TABLE 2
Human populations of the world's drylands

REGION	ARIDITY ZONE							
	Arid		Semi-arid		Dry subhumid		All drylands	
	(1 000)	(%)	(1 000)	(%)	(1 000)	(%)	(1 000)	(%)
Asia (incl. Russia)	161 554	5	625 411	18	657 899	19	1 444 906	42
Africa	40 503	6	117 647	18	109 370	17	267 563	41
Oceania	275	1	1 342	5	5 318	19	6 960	25
North America	6 257	2	41 013	16	12 030	5	59 323	25
South America	6 331	2	46 852	16	33 777	12	86 990	30
C. America and Caribbean	6 494	6	12 888	11	12 312	8	31 719	28
Europe	629	0	28 716	5	111 216	20	140 586	25
World total	222 043	4	873 871	4	941 922	17	2 038 047	37

Source: UNSO/UNDP, 1997.

in their crops or in their natural resource base. This leads not only to economic inefficiency but also to exploitation and degradation of the resource base, both of which in turn sustain their poverty. Evidence is often contradictory, for example, Mazzucato and Niemeijer (2000) found little evidence of widespread degradation of crop and fallow land discussing conditions in Burkina Faso. However. they do not dispute localized areas of severe degradation, nor suggest that Sahelian soils are particularly fertile. Mazzucato and Niemeijer (2000) do question the widespread belief that low-external-input practices used by West African farmers are leading to region wide land degradation and also argue that degradation assessments need to deal better with the spatial and temporal dimensions of the observed problems and land uses (Fresco and Kroonenberg, 1992; Rasmussen, 1999).

Dryland farming is dependent solely on the water available from precipitation. However, dryland farming often depends on having stored soil water at the time of seeding a crop to supplement the rainfall received during the growing season (Dregne and Willis, 1983). This is particularly true in Canada and the United States of America, also to some extent in Australia and Argentina, where dryland farming, at least until recently, depended largely on fallow conserving soil moisture. Thus, limited and high-risk production in one season is forfeited in anticipation that there will be at least partial compensation by increased crop production in the following season.

With the development of dryland farming in the steppe, water-conservation techniques had to be developed because moisture from seasonal precipitation was usually inadequate for crop growth and maturation (El-Swaify et al., 1985). The practice most widely adopted was summer fallow. Practised in areas around the Mediterranean for centuries, summer fallow became practical in the steppes and grew rapidly with mechanization. Summer fallow systems are inefficient because often only 15–20 percent of the precipitation occurring during the fallow period is actually saved as stored water. The remainder is lost through evaporation and runoff. However, summer fallow is very effective in reducing risk and ensuring some yield even in low-rainfall years (Lal and Pierce, 1991). Careful farming practices such as weed control, maintenance of a stubble mulch and leaving the surface in large clods can result in larger amounts of stored water.

Dryland farming is at best a risky enterprise. A favourable pattern of precipitation during the growing period can result in good yields even when the annual total is much below average. In contrast, there is no assurance of good production in years even when precipitation is greater than average if it occurs at times when crop water requirements are low.

LAND DEGRADATION IN DRYLANDS

The rapid population growth in drylands due to improvements in health conditions and other factors has placed tremendous pressure on the natural-resource base. Often, the inevitable result of increasing population in resource-poor areas is land degradation defined as the loss of production capacity of the land (FAO, 2000a). In this context, this should be understood as a result of a combination of natural processes and human activities that cause the land to become unable to properly sustain its ecological and economic functions. Traditionally, simplistic, explanations confused the issue,where the land users were considered the "guilty" - and blamed for the degradation (Gisladottir and Stocking, 2005). However, studies over the last ten to fifteen years conclude that the causes of land degradation are far more complex and that land degradation may result from much higher level policy and market failure rather than from failures of the poor land user (Gisladottir and Stocking, 2005). In the past drylands have been neglected by national and international development policies - adequate investments to generate positive results and reverse land degradation have been deficient (Johnson, Mayrand and Paquin, 2006).

Desertification (Plate 4) is defined as land degradation in drylands caused by climate variability and human activities (UNCCD, 2000; FAO, 2000a). In the case of drylands, land degradation results in desertification (Stewart and Robinson,1997). Droughts, common to these areas, exacerbate the degradation processes.

It is estimated that almost 75 percent of the cropland in Central America is seriously degraded; 20 percent in Africa (mostly pasture); and 11 percent in Asia (IFPRI, 2001). These data suggest that, globally, up to 40 percent of agricultural land is seriously affected by soil degradation.

Oldeman *et al.* (1991) suggest that land degradation can result in the following reductions in agricultural productivity:
- light: somewhat reduced agricultural productivity;
- moderate: greatly reduced agricultural productivity;
- strong: biotic functions largely destroyed, non-reclaimable at farm level;
- extreme: biotic functions fully destroyed, non-reclaimable.

Agricultural productivity is affected by many factors apart from soil quality, e.g. rainfall, deforestation, population pressures, climate, labour and technology. Because of the interdependent nature of land and its productivity, it is necessary to base claims of land degradation on multiple, complementary proxies that include properties of land (e.g. of soil, water and vegetation) as well as productivity indicators.

Tobler *et al.* (1995) stated that the increase in

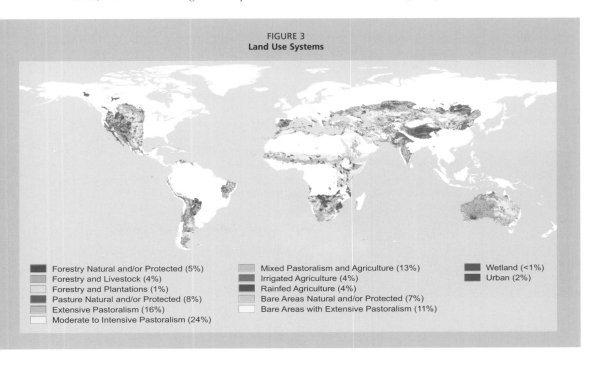

FIGURE 3
Land Use Systems

Forestry Natural and/or Protected (5%)
Forestry and Livestock (4%)
Forestry and Plantations (1%)
Pasture Natural and/or Protected (8%)
Extensive Pastoralism (16%)
Moderate to Intensive Pastoralism (24%)

Mixed Pastoralism and Agriculture (13%)
Irrigated Agriculture (4%)
Rainfed Agriculture (4%)
Bare Areas Natural and/or Protected (7%)
Bare Areas with Extensive Pastoralism (11%)

Wetland (<1%)
Urban (2%)

PLATE 4
Desertification in the Sudan. Goats feed on a solitary *Acacia* shrub (R. Faidutti)

population density is a major factor influencing land degradation, particularly in semi-arid and arid regions. Kirschke, Morgenroth and Franke (1999) claim that data from 73 developing countries have shown that deforestation is a causative factor for both wind and water erosion under arid and semi-arid conditions. Overexploitation of vegetation for domestic uses such as fuelwood and domestic timber is also a cause of degradation of the resource base, particularly in the Sahel belt of Africa, western Argentina, the Islamic Republic of Iran, and Pakistan (Kruska, Perry and Reid, 1995). Hazell (1998) stated that, despite some out-migration, human populations continue to grow in many less-favoured areas, but crop yields grow little or not at all, resulting in worsening poverty, food insecurity and widespread degradation of natural resources.

Another common problem associated with land degradation in drylands is the development of salt-affected soils (saline, saline–sodic and sodic soils) resulting from changes in the local water balance and the accumulation of excess salts in the rootzone. Dregne, Kassas and Rozanov (1991) estimated that about 41 million ha of irrigated land in the world's dry areas are affected by various processes of degradation, mainly waterlogging (20 million ha) and salinization and sodication (21 million ha). In the 11 countries surveyed with a total irrigated area of 158.7 million ha (70 percent of the world's irrigated land), 29.6 million ha (20 percent) are salt-affected soils.

Salinity also poses a major management problem in many non-irrigated areas where cropping relies on limited rainfall (FAO, 2005). Although dryland salinity has been a threat to land and water resources in several parts of the world, it is only in recent years that the seriousness of the problem has become widely known. In rainfed agriculture, intrusion of saline seawater to areas lying near the sea can cause land salinization during dry periods (Ghassemi *et al.*, 1995). Based on the FAO/UNESCO Soil Map of the World, the total area of saline soils is 397 million ha and that of sodic soils is 434 million ha that are not necessarily arable but cover all salt-affected lands at global level. If it is accepted that 45 million ha of the current 230 million ha of irrigated land are salt-affected soils (19.5 percent), then in the global total of the almost 1 500 million ha of dryland agriculture, 32 million ha are salt-affected soils (2.1 percent) caused to varying degrees by human-induced processes (Ghassemi, Jakeman and Nix, 1995). Salt-affected soils have lower productivity and need careful management but can be improved. Although many countries are using salt-affected soils because of their proximity to water resources and the absence of other environmental constraints, there is a clear need for a sound scientific basis to optimize their use, determine their potential, productivity and suitability for growing different crops, and identify appropriate integrated management practices.

All continental regions have experienced

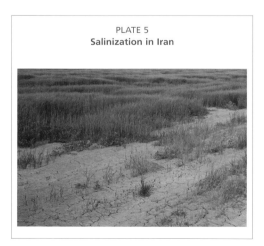

PLATE 5
Salinization in Iran

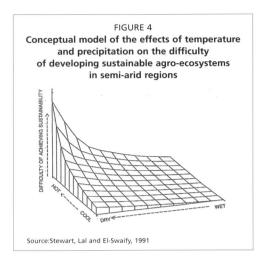

FIGURE 4
**Conceptual model of the effects of temperature
and precipitation on the difficulty
of developing sustainable agro-ecosystems
in semi-arid regions**

Source:Stewart, Lal and El-Swaify, 1991

a decrease in arable land per capita in every 5-year period reported. From 1965 to 1995, the decrease was 40 percent in Asia and more than 50 percent in Africa (FAO, 2000a). Sung-Chiao (1981) identified resource degradation as a major limitation to productivity in many arid and semi-arid regions of China.

In the past, under lower population density and with intensification of agriculture, drylands generally recovered following long droughts. However, under current conditions, they tend to lose their biological and economic productivity more rapidly and seriously. Stewart, Lal and El-Swaify (1991) developed a simplified conceptual model of the potential for soil degradation as the climate becomes hotter and drier. The relationships presented in Figure 4 suggest that soil degradation processes are more rapid in hotter and drier climates, making it more difficult to sustain the soil-resource base. Whenever an ecosystem such as a grassland prairie in a semi-arid region is transformed into an arable system for food and fibre production, several soil degradation processes are set in motion (Stewart, Lal and El-Swaify, 1991). This is particularly the case where raindrops fall directly onto the bare soil surface, not protected by vegetation, crop residues, mulches, etc. Other effects are a decline in soil organic matter (SOM), increased wind and water erosion, deterioration of soil structure, salinization and acidification.

EFFECTS OF LIVESTOCK ON THE RESOURCE BASE

Livestock have both positive and negative effects on the resource base, particularly the soil in drylands. They trample the soil which, depending on the soil type and status, frequently leads to compaction. The resulting higher bulk densities lower yields by inhibiting root development, reducing infiltration and waterholding capacity. Compaction also makes cultivation (mechanical or manual) more difficult and energy intensive.

Sandford (1988) reported that livestock, mainly cattle, consume up to 60 percent of the crop residues remaining on the surface after a grain harvest. Unless the manure is recycled, removal of crop residues by livestock (or for fuel uses) increases losses of nitrogen and phosphorus by 60–100 percent over the amounts removed in the grain. Where crop residues are burned on the land, most of the nitrogen and organic matter (on the surface and within the upper cms of the soil profile) are lost, but most of the other nutrients are recycled.

In some countries where dryland ecosystems predominate, overgrazing (Plate 6) is the major cause of land degradation, e.g. in the Libyan Arab Jamahiriya, Tunisia, the Islamic Republic of Iran, Iraq, the Syrian Arab Republic, and virtually the whole Sahel belt of Africa (FAO, 1998a). Overgrazing is also a major cause of land degradation in many parts of Central Asia, South America such as Brazil and Argentina, as well as in some developed countries including Australia and the western United States of America.

Drought presents major challenges for mixed farming operators in semi-arid regions. Mixed farming is defined here as involving crops, livestock and/or trees (ASA, 1976). Mixed farming systems include those in which more than 10 percent of the dry matter fed to animals comes from crop by-products/stubble or where more than 10 percent of the total value of production comes from non-livestock farming activities (FAO, 1996b).

In a drought situation, overgrazed mixed-farming systems can undergo the removal of much of the surface vegetation, increasing the soil's vulnerability to serious wind erosion (Dregne, 2002). A producer anticipating this

PLATE 6
The result of overgrazing on the outskirts of
Amman city (R. Faidutti)

problem should reduce the number of livestock, or remove them from the grazed areas and feed them a balanced ration for survival. Lot feeding and zero grazing become increasingly important strategies to retain plant cover and structural stability of the surface soil in order to sustain resources in the long term. These options are unlikely to be available to smallholders or pastoralists in developing countries.

With careful attention to the plant-nutrient balance, mixed farming is, environmentally, probably the most desirable system (Powell et al., 2004). Where appropriate, it should be the focus for farmers, agricultural planners and decision-makers. The challenge will be to identify the technologies and policies that enable sustained growth to satisfy the increasing demand for meat and milk. Despite the advantages of mixed farming, current trends in drylands point towards farmers choosing to specialise in either crop or livestock production. It is recommended that these trends should be resisted. It is being increasingly recognised that livestock play a vital role in stabilizing production and income in cereal-producing regions. Livestock graze crops/ residues and provide some income in years when climate conditions are so adverse that grain production is unprofitable. Livestock increase the sustainability of livelihoods in drylands, particularly in years of low precipitation.

EFFECTS OF CLIMATE CHANGE ON DRYLANDS

The global increase in atmospheric concentrations of carbon dioxide (CO_2) and other greenhouse

gases (particularly methane) over the past 250 years are attributed primarily to fossil fuel combustion and land use change (including *inter alia* deforestation, biomass burning, draining of wetlands, ploughing and use of fertilizers) and now far exceed pre-industrial values determined from ice cores spanning many thousands of years (IPCC, 2007). The global atmospheric concentration of carbon dioxide has increased from a pre-industrial value of about 280 ppm to 379 ppm in 2005 (IPCC, 2007). The atmospheric concentration of carbon dioxide in 2005 exceeds by far the natural range over the last 650,000 years (180 to 300 ppm) as determined from ice cores (IPCC, 2007).

There is substantial scientific evidence that the recent (and predicted future) rapid changes in the earth's climate are human-induced, caused by these accumulations of CO_2 and other greenhouse gases (GHGs) and have become a serious and urgent issue (Stern, 2006). Widely accepted predictions show that the on-going pattern of climate change will not only raise temperatures across the globe, but will also intensify the water cycle, reinforcing existing patterns of water scarcity and abundance, increasing the risk of droughts and floods. In addition, as the world warms, the risk of abrupt and large-scale changes in the climate system will rise – also the frequency and intensity of extreme events are likely to increase (Stern, 2007). The summary of the IPCC Fourth Assessment report (2007) states that hot extremes, heat waves, and heavy precipitation events will continue to become more frequent. Increases in the amount of precipitation are *very likely* in high latitudes, while decreases are *likely* in most subtropical land regions (by as much as about 20 percent in the A1B scenario in 2100), continuing observed patterns in recent trends. Tubiello and Fischer (2007), however, stated after taking into accout anticipated impacts of climate change and mitigation that in terms of cereal production, the impact on risk of hunger is only felt after 2050.

Climate change poses a real threat to the developing world which, unchecked, will become a major obstacle to continued poverty reduction (Stern, 2006). Developing countries are especially vulnerable to climate change because of their geographic exposure, low incomes, and greater reliance on climate sensitive sectors such as agriculture (Stern, 2006). The impacts of climate change in drylands are likely to lead to still more people and larger areas of land being affected by water scarcity and the risk of declining crop yields – with the peoples of drylands in developing countries least able to adapt due to poverty.

Adoption of any strategy(ies) to increase agricultural production in drylands provides a much needed route to adapt to the effects of the current period of rapid climate change, which otherwise will become a major obstacle to poverty reduction. The options outlined in Chapter 4, offer approaches by which smallholders and other land users in drylands can adapt to cope with changing climate, including improved *in situ* water conservation, water harvesting and reduction in evaporation. Improved management of soil organic matter and conservation agriculture will not only help small holders and pastoralists adapt to climate change, but also involve changing traditional agricultural practices to increase storage of C in soils and on the soil surface, contributing to mitigating emissions of GHGs. Shifting agricultural zones, planting of drought resistant/fast maturing strains of crops and protection of local agro-biodiversity offer other ways by which smallholders and pastoralists can cope with the rapid rate of human-induced climate change.

CHAPTER 2

Cereal production in Drylands

RECENT TRENDS IN WORLDWIDE CEREAL PRODUCTION

The world's food supply is obtained either directly or indirectly from plants, but fewer than 100 are used for food (Burger, 1981). Worldwide, about 50 species are cultivated actively, and as few as 17 species provide 90 percent of world's food supply and occupy about 75 percent of the total tilled land on earth (Harlan, 1976). The important plant species include wheat (*Triticum aestivum*), rice (*Oryza sativa L.*), maize (*Zea mays L.*), potato (*Solanum tuberosum*), barley (*Hordeum vulgare*), sweet potato (*Ipomea batatas*), cassava (*Manihot esculenta*), soybean (*Glycine max*), oat (*Avena sativa*), sorghum (*Sorghum bicolor*), millet (*Pennisetum typoides*), rye (*Secale cereale*), peanut (*Arachis hypogaea*), field bean (*Dolichos lablab var. purpureus*), pea (*Pisum sativum*), banana (*Musa paradistaca*), and coconut (*Cocos nucifera*).

Eight cereal grains — wheat, barley, oat, rye, rice, maize, sorghum and millet provide 56 percent of the food energy and 50 percent of the protein consumed on earth (Stoskopf, 1985). Cereals continue as the most important source of total food consumption in the developing countries where direct consumption of grains provides 54 percent of total calories and 50 percent for the world as a whole (FAO, 2006). Wheat and rice are by far the most widely consumed cereals in the world. Maize is a major crop for both direct and indirect human consumption because it is a major energy feed for animals. Wheat, rice, and maize constitute approximately 85 percent of the world's production of cereals.

Cereal crops are grown in countries across the globe. One or more cereal grains are basic crops in the seven major areas of the world — Africa, North America, South America, Asia, Europe, Oceania, and the former USSR (Stoskopf, 1985). Adapted cereal crop species and cultivars within each species are found in all latitudes from 60' N to 50' S, on soils ranging from slightly acid to alkaline, in both arid and well-watered regions. Only rice is slightly more confined to low/ middle latitudes. Rice is mostly grown under irrigated conditions and maize is usually limited to irrigated areas or regions where precipitation is both adequate and dependable. Wheat is the most widely grown cereal crop and is extensively grown in dryland regions under both non- irrigated and irrigated conditions. Every month of the year, a crop of wheat is being harvested somewhere in the world, from as far south as Argentina to as far north as Finland. Wheat is best suited to areas between 30' and 50' N latitude and 25' and 40' S latitude. Wheat is a major crop of the United States and Canada, Australia and is grown extensively in almost every country in Latin America, Europe and Asia. Well defined environmental conditions must be met, however, with respect to temperature, precipitation, frost-free period, and soil if wheat production is to be successful. Winter wheat crops usually produce higher yields than spring wheat and are more extensively grown. Two factors limit the growing of winter wheat — the ability to overwinter high latitudes and a vernalization or cold temperature-photoperiod interaction at low latitudes (Stoskopf, 1985). Maize can be grown as far north as the 50th parallel in Canada and over most of the United States, throughout Mexico and Central America, to as far south as central Argentina and Chile (about 35' S latitude) in South America (Stoskopf, 1985). Maize is also adapted to Africa, central Europe and Asia, making it a crop that is universal in its adaptation.

Production data for cereals in the major world regions – in the period 1961 to 2006 are presented in Annex 3, Table 1. Those figures and Table 3 show that world population of cereals increased 2.53 times in that period. In 2006, wheat, rice and maize accounted for more than 87 percent of cereal production, compared with 73 percent in 1961. Maize has shown a rapid increase in recent years and accounted for 31 percent of the cereal production in 2006 compared to only 23 percent in 1961. This percentage will almost certainly increase further in future years, as more maize is used for production of biofuel. In terms of food security, a very positive fact is that cereal production has increased at a faster rate in the developing countries than in the developed countries. From 1961 to 2006, cereal production in developing countries increased 2.7 times compared to 2.3 times for the developed countries. The production in developing countries increased from about 190 kg/capita in 1961 to about 250 in 1985, but has remained almost constant for the last 20 years. For the entire world population, cereal production per capita increased from 286 kg in 1961 to 371 kg in 1990, and has decreased slightly since 1990 (Table 3). The impact of increasing amounts of maize (and other cereal crops) being diverted from food to industrial uses, creates some uncertainties for the future.

Between 1961 and 1994, wheat yields increased at an average annual rate of about 2 percent in developing countries except China and India, the two largest wheat producers (Pingali and Rajaram, 1999). In the Near East/North African countries and the wheat-producing countries of sub-Saharan Africa, wheat yields grew at about 2.4 percent/year from 1961 to 1994. This compared with about 1.8 percent for Latin America, which started at a higher base. Yields in India rose sharply in the early years of the Green Revolution, from the mid-1960s until the late 1970s. In China, grain yields rose rapidly after rural reforms began in 1978. Between 1978 and 1990, wheat yields increased from 1.8 to 3.2

TABLE 3
Trends in world production of cereals

	1961	1970	1980	1990	2000	2006
	(million tonnes)					
Wheat	222	311	440	592	580	606
Rice	216	316	397	520	597	635
Maize	205	266	397	483	589	695
Sorghum	41	56	57	57	59	56
Millet	26	33	25	30	27	32
Total cereals *	877	1 193	1 550	1 953	2 051	2221
Cereals/capita (kg)	286	323	349	371	339	336
Used for feed (%)	31	37	37	33	33	
Arable land (million ha)	1 266	1 302	1 331	1 383	1 380	
Irrigated land (million ha)	139	168	210	244	271	
Fertilizer used (million tonnes)	31	69	117	138	137	

* Totals rounded independently

Source: FAOSTAT (www.fao.org)

tonnes/ha, maize yields from 2.8 to 4.5 tonnes/ha and paddy rice yields from 3.9 to 5.7 tonnes/ha. In 2002, Chinese grain yields for wheat, maize and paddy rice were 3.9, 5.0 and 6.3 tonnes/ha, respectively (FAO, 2003c).

The rise in grain production, particularly in developing countries between the 1960s and 1990s was largely a result of increased irrigation and inorganic fertilizer use, along with better crop varieties, pest control, and other management improvements. Between 1961 and 2000, the irrigated area in developing countries more than doubled. The amounts of fertilizer used increased 24 fold in developing countries and 34 fold in the developing countries in Asia (FAOSTAT, 2007). Fertilizer use in developed countries doubled during the same period. As a group, developing countries have been narrowing the gap in grain production with developed countries.

Despite the increase in cereal production in developing countries between the 1960s and the 1990s, in recent years, a decrease in the major producing countries has been recorded. For example in China, the production of grain, particularly wheat, has declined from 123 million tonnes in 1997/98 to 100 million tonnes in 2000/01 (Hsu, Lohmar and Gale,

2001) with a production of 91 million tonnes being recorded in 2004 (FAOSTAT, 2007). The decline is attributed to a combination of drought, oversupply, reduced government support, shift to higher-income crops and a shift in emphasis from quantity to quality. China's production of wheat, maize and rice fell by a combined total of 32 million tonnes in 2004 compared with 1999 (FAOSTAT, 2007).

While global food demand is increasing, the agricultural resources for producing grains are being diverted increasingly from cereal production to other agricultural and non-agricultural activities. Pingali and Rajaram (1999) forecast that the global demand for wheat in 2020 will be 40 percent greater than the 552 million tonnes produced in 1997, mainly because of increased demand in developing countries. Pingali and Pandey (2000) projected that global maize demand in 2020 will be 50 percent greater than the 558 million tonnes produced in 1995. The increased demand for maize in recent years has largely been driven by rising incomes and the consequent growth in meat and poultry consumption.

A paradigm shift to the use of cereals for producing ethanol for fuel is presently occurring, which could have far-reaching consequences. Maize in the U.S. and wheat in Europe are being used in rapidly increasing quantities for producing ethanol for fuel as a result of changes in the cost and availability of oil, concerns over fuel supply, fuel security and Kyoto Protocol compliance. Although it is too early to fully determine the impact of this, the fact that food systems and energy systems are competing for cereals is a historic development. The U.S. produces about 40 percent of the world's corn and has usually exported about 20 percent of this accounting for about 60 to 70 percent of the exports from all nations. However, ethanol production is increasing so rapidly in the U.S. that exports will almost certainly be reduced significantly and the price of exported maize is likely to increase significantly. The Institute for Agriculture and Trade Policy (2006) reported that ethanol production consumed less than 5 percent of the U.S. corn crop 10 years ago. In contrast, ethanol production consumed about 14 percent of the 2005 crop, and estimates indicate that it was more than 20 percent in 2006. Furthermore, an additional 58 plants are currently under construction or being expanded to supplement the more than 100 ethanol production plants. Europe and China are also using cereals as feedstock for ethanol fuel. The competition for maize and other cereals has resulted in an increase in price, and while this will result in increased production of cereals, the amount used for food could decrease and this could decrease meat production and the amounts of maize available for exports – raising issues of food security. In December, 2006, the National Development and Reform Commission of China ordered local governments to stop approving new projects that process maize for industrial uses. China used more than 23 million tonnes of maize for industrial uses in 2005, an increase of 84 percent from 2001, while production of maize increased only 21.9 percent over the same period (People's Daily Online, 2006). The population in many dryland areas is increasing significantly and it will become increasingly important that these areas are as self-sufficient in cereal production as feasible. Developing countries must either produce sufficient grain for their needs or produce enough foreign exchange to import their needs. The escalating use of grain for fuel production will almost certainly result in smaller amounts available for import and higher costs.

Pingali and Rajaram (1999) claim serious concerns about future wheat supplies have emerged for the first time since the start of the Green Revolution. Newer generations of HYVs continue to improve on yields of earlier varieties. However, starting in the 1990s, the rate of growth in yield potential has slowed considerably. In environments favourable for wheat production, the economically exploitable gap between potential yields and farmer yields has been reduced considerably in the past three decades. Therefore, given existing technologies and policies, the cost of marginal increments in yield could exceed the incremental gain. In addition, decades of poor water management have caused large tracts of irrigated land to be abandoned or cultivated at lower levels of productivity due to salinization (Ghassemi et al., 1995). The Indian Punjab, Pakistani Punjab, the Yaqui Valley in northwest Mexico and the irrigated wheat areas in the Nile Valley all exhibit visible signs of land degradation attributable to salinity build-up. Salinity is thought to affect nearly 10 million ha of wheat in developing countries (Ghassemi et al., 1995).

THE ROLE OF LIVESTOCK IN CEREAL PRODUCTION

It is forecast that livestock will play an increasingly important role in cereal-producing regions in arid and semi-arid regions in future. In addition to consuming cereal grains, livestock utilize large tracts of land by grazing and foraging in these pastoral/agropastoral systems.

Global livestock numbers are increasing rapidly. Between 1970 and 1990, annual meat consumption per capita in developing countries increased from 11 to 18 kg, and by 2000, it had reached almost 27 kg (FAO, 2006). Milk consumption in developing countries increased from 29 kg per capita in 1970 to 38 kg in 1990 and 45 kg in 2000. Although these gains are very significant, they are still far below the 90 kg meat and 214 kg milk per capita in the industrial countries (FAO, 2006). The growth in demand for meat and milk products will continue, particularly in those developing countries where incomes and living standards are improving. The effect of this on cereal-growing areas will vary depending on the region. The average proportion of total cash income derived from livestock is much higher in semi-arid and subhumid regions than in more humid regions, where crop production is the principal income source. Sandford (1988) reported that 40 percent of Africa's human agricultural population and an estimated 57 percent of the ruminant livestock live in the semi-arid and subhumid regions of sub-Saharan Africa. Income from livestock in these drier regions accounted for more than half of farmer incomes, compared with less than 10 percent in the humid regions. Livestock systems are usually less variable livelihood sources than grain systems, and they also provide opportunities for increasing the use efficiency of limited water resources (Srivastava et al., 1993). This assertion is supported by Sandford (1988), who reported that the variations in annual livestock output in the Ethiopian highlands were ±10 percent compared with ±32 percent for grain production. This evidence indicates that mixed-farming systems can be expected to be considerably more sustainable than purely arable systems in dryland regions.

Mann (1991) reported that the most successful integration of crops and livestock in Australia has been in the cereal–livestock zone in the south of the country. The pasture phase is an important component of the system, and promotes high levels of both crop and livestock production where managed properly. Crop residues are available for livestock and careful grazing can help prepare the land for the following crop, while at the same time maintaining a suitable amount of cover for protection of the land. Although integration brings benefits to both crops and livestock, there are some constraints that may reduce flexibility in both cropping and livestock operations.

THE ROLE OF IRRIGATION IN CEREAL PRODUCTION

There is a water crisis today, not because of having too little to satisfy the needs of the population, but because of the difficulty of managing water so that people and the environment do not suffer (Cosgrove and Rijsberman, 2000). Twenty percent of the world's population do not have access to safe and affordable drinking-water. Even though people use only a small fraction of

renewable water resources globally, this fraction is much higher (up to 80–90 percent) in many arid and semi-arid river basins where water is scarce (FAO, 1996a). As populations increase in semi-arid regions, as is the case in many developing countries, water requirements for industry and domestic use are increasing at the expense of irrigation needs. As people in dryland regions improve their living standards, the competition for water will accelerate. The World Bank (2000) reported that the share of extracted water used for agriculture ranged from 87 percent in low-income countries, through 74 percent in middle-income countries, to 30 percent in high-income countries.

In the future, the way in which water is managed will have a dramatic effect on irrigation and thus on food production. Cosgrove and Rijsberman (2000) consider that a reduction in the rate of expansion of irrigated agriculture is crucial to deal with the water crisis. They call for a 40 percent increase in food production by 2025 with only 9 percent more irrigation water. This would require that the increase in food production comes, to a great extent, from non-irrigated agriculture.

Water for irrigation expansion is becoming harder to find and more costly to develop. At the same time, a proportion of currently irrigated land is threatened because of soil degradation, particularly the buildup of salts and depletion of water resources. For example, the Ogallala aquifer in the Great Plains of the United States of America supplies water for more than one-quarter of that country's irrigated area. In less than 50 years of pumping, many of the wells, particularly in the southern portion of the aquifer, have become so unproductive that the land has been returned to dryland management. Groundwater depletion is also a major problem in central and northern China, north-western and southern India, parts of Pakistan, much of the western United States of America, North Africa and the Near East. Postel (1999) concluded that unsustainable exploitation of groundwater might have become the single largest threat to irrigated agriculture, exceeding concerns over the build-up of salts in the soil.

Hsu, Lohmar and Gale (2001) discussed the effect of reducing water resources gradually on wheat production in China. Irrigation has been a very important factor in the rapid growth of grain production in China. However, a lack of irrigation water in the future is a major constraint. The groundwater tables in the important wheat-producing provinces of Henan, Shandong and Hebei have been falling rapidly. More than half of the irrigation water supplies in Hebei and more than 40 percent in Shandong are from groundwater. Many rivers and streams in the area have also been overexploited and are dry for much of the year. In this region, wheat is heavily dependent on irrigation because the main growing period is in the dry spring season (Hsu, Lohmar and Gale, 2001).

Globally, the development of irrigated land has been slowing. During the 1960s, irrigation area expanded 2.1 percent per year and reached a peak of 2.2 percent during the 1970s. The rate slowed to 1.6 percent during the 1980s and to 1.2 percent in the 1990s, and the rate from 2001 to 2003 was only 0.1 percent (FAO, 2008). Postel (1999) projected that the global irrigation base is unlikely to grow faster than 0.6 percent/year in the next 25 years, and that this figure may still turn out to be optimistic. Svendsen and Turral (2007) reported that there were 208.7 million hectares of irrigated land in developing countries in 2002 and 68.1 million in developed countries. Molden *et al.* (2007) showed that while the world's cultivated land increased from 1 368 million hectares to 1 541 between 1961 and 2003, irrigated area almost doubled from 139 million hectares to 277 million. They showed, however, a marked decline in expansion during the past few years. Globally, donor spending on irrigation peaked in the late 1970s and early 1980s and then fell to less than half (Molden *et al.*, 2007). They stated that four factors contributed to the decline. First, there was a sharp drop in cereal prices. Second, there was growing recognition of poor performance of irrigation systems. Third, there was a rise in construction costs of irrigation infrastructure. Fourth, there was growing opposition to environmental degradation and social dislocation often associated with large dams. Recently, there has been renewed interest by the World Bank for reengaging in agricultural water management (World Bank, 2006). The sharp increases in cereal prices since 2005 could also result in renewed interest. World population growth has also slowed in recent years, but per capita irrigation area peaked in 1978 and has fallen 5 percent since then.

It is estimated that irrigated land which to date cover 197 million ha in developing countries, will increase by 45 million ha by 2030 (FAO, 2002). Most of this increase will be achieved by providing irrigation to rainfed land or from land with rainfed potential not yet in use. Of the area currently irrigated, it is estimated that 42 million ha are in arid and hyper-arid climates, and that 5 million ha of the projected increase will be in such regions. The expansion of irrigation will be greatest in countries that have few land reserves and are hard-pressed to raise crop production through more intensive cultivation practices such as in South Asia, East Asia, and the Near East and North Africa. Four countries (India, China, the United States of America, and Pakistan) account for more than half of the world's current irrigated land. Ten countries, including the Islamic Republic of Iran, Mexico, Russia, Thailand, Indonesia and Turkey, account for two-thirds of the world's irrigated land (Postel, 1999). FAO (2003b) predicts an average increase of 0.6 percent a year between 1997/99 and 2030 in developing countries, compared with 1.6 percent a year from 1960 to 1990. Svendsen and Turral (2007) state that this will still result in an increase to 45 percent of agricultural production coming from irrigated land by 2030. They also report that this will mean that the amount of water withdrawn for irrigation will increase by 12 to 17 percent above the present level.

Although the costs of developing land for irrigation vary widely both within and among countries, they are becoming increasingly difficult to justify. This is particularly true for cereal production. Future irrigation development will probably become increasingly limited to high-value crops such as vegetables, fruits, and tree crops. It is becoming more important to invest in dryland soil- and water-conservation practices that can increase grain production in dryland regions. Many of these areas already have existing and growing populations whose demands for grain exceed their production, and grain imports are essential. Although increases in dryland grain production may not be able to eliminate the need for imports, the amounts can be reduced, while at the same time the increased production can improve the economic conditions and living standards of the people. An FAO study (FAO, 1998b) considered that it would be an error to disregard the potential to increase food production from dryland agriculture because of the difficulties associated with it; where dryland agriculture is inefficient, there is scope for increasing food production by improvement. Improving dryland agriculture also shortens the food cycle and enhances food security by producing the food where the consumers are located.

EXPANSION OF CEREAL PRODUCTION

Borlaug (1996a) suggested that improvements in overall crop management can still increase yields by 50–100 percent in much of South and Southeast Asia, Latin America, the Commonwealth of Independent States and Eastern Europe, and by 100–200 percent in most of sub-Saharan Africa. In the case of China, the United States of America, and the European Union, where yields are already high, it will be difficult to achieve further increases. According to Borlaug (1996a), the last major land areas for developing cropland are the acid soils of the Brazilian cerrado, of the Colombian and Venezuelan llanos, and of central and southern Africa.

The central cerrado of Brazil, with about 100 million ha considered potentially arable, is the single largest contiguous block of uncultivated land that can contribute to world food production in the next three decades. This could increase the world's arable land by about 7 percent. However, bringing these potentially arable lands into cultivation presents formidable challenges, and sustaining their productivity may be even more difficult. The soils of this area are mostly deep loam to clay-loam Ferralsols and Acrisols (Oxisols and Ultisols) with good physical properties, but highly leached of nutrients (Borlaug, 1996b). They are strongly acid, with toxic levels of soluble aluminium and manganese; most of the soil phosphate is fixed and unavailable to plants (Furley and Ratter, 1988; Sanchez, 1997). However, there are some varieties of wheat, maize, soybeans, rice, triticale and several species of pasture grass with aluminium tolerance. The degree to which crop production on these soils can be sustained is still a matter of

conjecture and further research.

The potential increase in arable lands in South America will be offset partially by losses in South Asia. Borlaug (1996b) suggests that 21 million ha are being cultivated in South Asia that should not be. These lands are either too arid or so vulnerable to erosion because of topography that they should be removed from cultivation. China has reported that there are 15 million ha of cropland in China with slopes greater than 25 degrees (almost 50 percent), with 70 percent of these lands in the west (China Ministry of Science and Technology, 2001). China began removing these lands from cultivation in Sichuan, Shaanxi and Gansu provinces in 1999. Pilot studies are underway to reclaim lands in 224 counties in 20 provinces and autonomous regions and municipalities in the central and western parts of China.

The Conservation Reserve Program (CRP) was established in the United States of America as part of the 1985 Food Security Act. Under the act, grass has been seeded or trees have been planted on about 14 million ha of highly erodible cropland. Although some of this land may be returned to crop production in the future, the CRP will probably result in a permanent reduction in cropland area.

In view of the probability that there will not be a large overall increase in the amount of arable land to meet the growing demand for cereal grains, yields will have to continue to increase significantly to meet demands not only for food but also for the range of feedstocks for biofuels. Grain yields are determined primarily by soil fertility and water. The 225 percent increase in cereal yields since 1961 has been due largely to a doubling of the extent of irrigated land and the more than fourfold increase in fertilizer use (Annex 3, Table 1). It is unlikely that these inputs can continue to grow at such a rate as, in order to meet future cereal demands, irrigated land would need to expand 20–30 percent by 2025. To meet this, irrigation-water usage would have to increase at least 17 percent above the 1995 level by 2025 (Shiklomanov, 1999; Seckler et al., 1998), even using optimistic assumptions on yield and efficiency improvements. A 30 percent increase in irrigated area would require major investments in water infrastructure, including large dams. This would probably result in severe water scarcities and risk serious deterioration of ecosystems. However, in contrast, a major reduction in the rate of expansion of irrigation could possibly lead to food shortages and rising food prices. There is already evidence that the rate of irrigation expansion is slowing. According to Rosegrant and Ringler (1999), the annual growth rate in global irrigated area declined from 2.2 percent between 1967 and 1982 to 1.5 percent between 1982 and 1993.

Additional increases in cereal production in developing countries may come from a combination of three approaches:
- The continuation of the development of irrigated lands coupled with HYVs, fertilizers, pesticides and other inputs. This approach has been extremely successful in the past few decades but is becoming increasingly difficult and expensive.
- The development of new areas for rainfed crop production. The FAO model for land resource potential (FAO, 2001a) indicates that there is considerable potential for additional cultivatable land particularly in sub-Saharan Africa and South and Central America. According to the model, 133 million ha of land in Africa are very suitable for rainfed cereal production, and another 558 million ha are suitable or moderately suitable. However, in 2000, only 340 million

ha were under maize and wheat cultivation. Thus, there are many constraints on cereal production in Africa other than soil and water resources. These constraints relate mostly to poor institutional, infrastructural and financial capacities of African countries.
- The improvement of soil- and water-management practices on existing rainfed lands. FAO (2007b) recently published a land productivity potential for agriculture particularly classifying the suitability of rainfed cereal production. The suitability of global land area for rainfed production of cereals with various levels of inputs: low, intermediate and high levels of inputs as well as variability in rainfed production is contained in Annex 4.

If more cereal grains can be produced with the same amount of water (or less) by better water management, water harvesting and water conservation, local food security will be enhanced and there will be less competition for water. The corollary is that more water will remain for household and industrial uses, also to sustain vital ecosystem functions. It is clear that the water-use efficiency of both irrigated and non-irrigated agriculture must be increased.

The following chapters focus on improving soil- and water-conservation practices to increase water-use efficiency in drylands.

CHAPTER 3

Enhancing cereal production in Drylands

Grain production in dryland areas must be improved to help meet the requirements of a growing world population, urbanisation and the transition to meat-rich diets. A major contribution to this improvement will be the capture and use of a greater portion of the limited and highly variable precipitation in dryland areas. Several approaches and practices of water and soil conservation and management can increase water-use efficiency, thus increasing yields and reducing the likelihood of crop failure.

One of the difficulties with crop production in dryland regions is the extreme variation in precipitation and, therefore, yields between years. Annual precipitation in dryland regions commonly ranges from less than half of average in a dry year to more than twice average in a wet year, which renders the use of averages of little use in planning agricultural and natural resource development. Ephemeral streams are the norm rather than perennial streams and it is not uncommon for perennial streams originating from higher elevations (orographic rainfall) to become intermittent downstream. As a consequence of the highly variable annual precipitation, yields can vary from zero to about three times average. Much of the precipitation in dryland regions occurs during high-intensity localized convective storms which result in a high level of spatial variation in rainfall – flash floods can transform a dry channel into a torrent in hours resulting in significant runoff while a channel a few km away may be completely dry (Brooks and Tayaa, 2002). The development and implementation of technologies that reduce the losses following rainfall events, including flash floods, could increase crop yields considerably.

Cereal yields in drylands could be increased and the year-to-year yield variability could be reduced if substantial effort and capital were invested. Investment in dryland agriculture represents a more environmentally sustainable and lower -cost alternative to the large amounts of capital being used to develop additional irrigated lands. Examples of the cost of developing additional irrigated lands are: US$8 300/ha for sub-Saharan Africa, US$6 700/ha for North Africa and the Near East (FAO, 1995), and US$12 300/ha for Baluchistan Province in Pakistan (Venkataraman, 1999). More recent irrigation development costs are available at AQUASTAT, 2008. In comparison, the costs of dryland improvement by water harvesting are in the order of US$300 to 5 000/ha (Oweis, Prinz and Hachum, 2001). While this improved cropland still would not be as reliable as irrigated land for cereal production because of droughts, such an investment could improve long-term average yields and yield stability significantly.

An important option for grain and other agricultural production in drylands is water harvesting. Water harvesting, which includes runoff farming, runoff storage and dry farming using fallow storage, can be less costly than irrigation and can be developed locally depending on rainfall and land conditions (Ben-Asher, 1988; Reij et al., 1988); FAO, 1991; Suleman et al., 1995).
A major difference between irrigation and water harvesting is the farmer's control over timing. Water can only be harvested when there is precipitation, so there is no assurance against crop failure in years when the precipitation is so low that there is little or no runoff or storage. This lack of security in water-harvesting schemes

compared with irrigated land is one reason why lending institutions and development organizations have been reluctant or unwilling to invest in water-harvesting schemes. However, the time may be right to rethink investment in water-harvesting practices. There is sufficient rainfall and soils information in most dryland regions, coupled with models that can analyse and determine probabilities, to design water-harvesting schemes that will improve crop production in the majority of years. These schemes may be more cost-effective than developing additional irrigated lands where water resources are limited. More importantly, these schemes can be developed in areas where there is no water available for irrigation.

WATER-USE EFFICIENCY

Water-use efficiency is an important concept for understanding soil–crop systems and designing practices for water conservation (Cooper et al., 1987; Howell, 1990; Musick et al., 1994; Rockström, 2000; Australian Centre for International Agricultural Research, 2002). It is defined as the amount of harvestable product produced per unit of evapotranspiration between the dates when the crop is seeded and harvested, commonly expressed in kilograms per cubic metre. Evapotranspiration is the sum of the amounts of water transpired by the crop and lost by evaporation from the soil surface.

Biomass production, grain yield, transpiration and evapotranspiration are related and are contributing factors to water-use efficiency. In years of below-average precipitation, the threshold amount of evapotranspiration may not be met or only exceeded by a small amount; therefore, little or no grain is produced. Just a small amount of additional water can increase yields dramatically once the threshold amount has been reached. For example, sorghum grown in semi-arid regions requires about 100 mm of seasonal evapotranspiration before any grain is produced (Stewart and Steiner, 1990). About 15 kg/ha of sorghum grain can be produced for every additional millimetre of evapotranspiration.

Data from Bushland, Texas, in the semi-arid Great Plains of the United States of America, for grain sorghum (Stewart and Steiner, 1990), maize

(Howell, 1998), and wheat (Musick *et al.*, 1994) are summarized and compared in Figure 5. Grain sorghum has a yield advantage over maize where seasonal evapotranspiration is limited because it requires less water to initiate grain production. However, once the threshold value for initiating grain production has been met, maize produces more grain for each additional unit of water than either grain sorghum or wheat. The relationships shown in Figure 5 indicate why maize is usually the crop of choice under favourable water conditions and why grain sorghum performs best when water resources are limited. An understanding of these relationships coupled with information about the probabilities of seasonal precipitation and the amount of stored plant-available water in the rootzone at seeding time allows producers to assess production risk.

Stewart, Jones and Unger (1993) compared annual cropping of wheat at three semi-arid locations in Australia, China and the United States of America (Annex 3, Table 3). The percentage of total precipitation used for evapotranspiration was similar for all three locations at about 65 percent. In all locations, plant-available water decreased during the growing season and increased during the fallow period. However, the change was considerably less for the location in Texas in the United States of America. This had less precipitation during the fallow period, and a very high potential evapotranspiration; it is

the most arid of the three locations. While total precipitation was greater at the Texas site than at the China site, actual evapotranspiration during the wheat-growing season was also greater. The site in China had a much higher yield, and a water-use efficiency of 0.47 kg/m³ compared with 0.31 kg/m³ for the Texas location. Water-use efficiency values for wheat grown in humid regions or under irrigation often exceed 1.25 kg/m³ and values as high as 1.9 kg/m³ are reported in the literature (Musick and Porter, 1990).

Figure 6 shows the relationships between grain yield of wheat and seasonal water use for the sites in Texas (United States of America) and China. At both locations, about 200 mm of evapotranspiration were required before any grain was produced. However, for each additional millimetre of water use, about 12 kg/ha of grain was produced at the Texas site compared with about 25 kg/ha at the China site. The relationships shown in Figure 7 illustrate the impact of technologies that increase the amount of water available for crop use and the resulting grain yield, but the degree of impact will be site-specific.

Increased soil-water storage also influences the effects of fertilizer and other inputs. FAO (2000b) developed a generalized relationship between water use and cereal grain yields showing that the impact of inputs increased sharply with

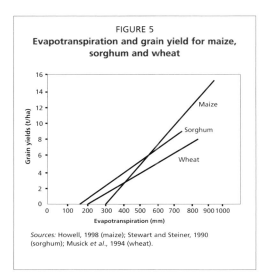

FIGURE 5
Evapotranspiration and grain yield for maize, sorghum and wheat

Sources: Howell, 1998 (maize); Stewart and Steiner, 1990 (sorghum); Musick *et al.*, 1994 (wheat).

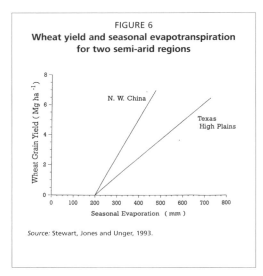

FIGURE 6
Wheat yield and seasonal evapotranspiration for two semi-arid regions

Source: Stewart, Jones and Unger, 1993.

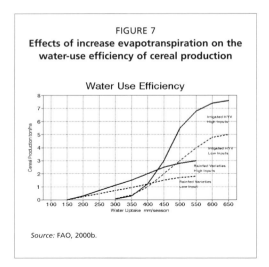

FIGURE 7
Effects of increase evapotranspiration on the water-use efficiency of cereal production

Source: FAO, 2000b.

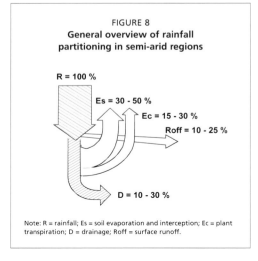

FIGURE 8
General overview of rainfall partitioning in semi-arid regions

Note: R = rainfall; Es = soil evaporation and interception; Ec = plant transpiration; D = drainage; Roff = surface runoff.

increased water availability (Figure 7). There are crop models that estimate grain yields for various cereals based on average and predicted amounts of evapotranspiration (FAO, 2003c). While not necessarily accurate for a specific year, these models are extremely useful in assessing the suitability of an area over a number of years.

The fact that cereal yields tend to increase in proportion to increases in evapotranspiration makes it imperative that dryland farming systems focus on reducing runoff and evaporation from the soil surface (Lal and Pierce, 1991). This allows a higher proportion of the limited precipitation to be used for transpiration, leading directly to higher yields. This is even more important where fertilizer and other inputs are used to allow a higher nutrient-limited yield.

The major emphasis in dryland farming is to capture, store and utilize highly variable and scarce precipitation. An overview of the partitioning of rainfall in the semi-arid tropics is shown in Figure 8. A large proportion of non-productive water flow in the dryland crop-water balance indicates problems that may be related to soil-fertility depletion or soil physical deterioration (especially reduced infiltration and waterholding capacity) through oxidation of organic matter. As much as 70 percent of precipitation may not be used directly for crop production. The focus must be to use more of the water as transpiration and lose less to runoff

and evaporation from the soil surface (Cooper *et al.*, 1987); Stewart and Steiner, 1990; Howell, 1990; Musick *et al.*, 1994; Rockström, 2000; Australian Centre for International Agricultural Resreach, 2002). This can be done by two different management strategies: *in situ* water conservation; and water harvesting.

In situ water conservation aims to prevent runoff and keep as much rainfall as possible where it falls, and then minimize evaporation, so that the water remains available for the crop. Water harvesting (runoff agriculture and runoff storage) is the collection and concentration of rainwater and runoff and its productive use for crops, livestock or domestic use (Smith and Critchley, 1983; Reij *et al.*, 1988; Oweis *et al.*, 2001). Water-harvesting practices are often designed to enhance runoff in one area so that the water can be used by a crop on an adjacent area or stored and used later, often at another site.

IN SITU WATER CONSERVATION

The concept of "blue" and "green" water is used to distinguish between two fundamentally different elements of the water cycle (Falkenmark, 1995). After atmospheric precipitation reaches the land surface, it divides into different sections which pursue the terrestrial part of the hydrological cycle along different paths. UNDP (2006) estimates that of the 110 000 km³ of precipitation falling annually on the land surface, about 40 000 km³ is converted into surface runoff and aquifer recharge (blue water) and about 70 000 km³ is stored in the soil and returned to the atmosphere through evaporation and transpiration by plants (green water). Rainfed agriculture uses only green water, while irrigated agriculture uses blue water in addition to green water. UNDP (2006) estimated that crop production uses up to 13 percent (9 000 km³/yr) of the green water while the remaining 87 percent is used by the non-domesticated vegetation including forests and rangelands. It is further estimated that about 2 300 km³/yr is withdrawn from rivers and aquifers for irrigation, but only about 900 km³ is effectively consumed by crops. Molden *et al.* (2007) states that 80 percent of agricultural evapotranspiration is directly from green water, with the rest from blue water sources.

Controlling runoff is a primary objective of any dryland cropping system. Although total precipitation in dryland regions is limiting, high-intensity storms are common and the amounts of runoff ("blue" water – Falkenmark, 1995) can be significant. The amount of runoff is often largely independent of slope. Rockwood and Lal (1974) reported about 20 percent runoff from bare fallow land in Nigeria regardless of whether the slope was 1, 5, 10 or 15 percent. While the runoff amounts were similar, the amount of erosion was strongly dependent on the slope.

The potential for runoff increases with a decline in SOM because soil structure deteriorates and surface crusts form. Runoff can be particularly high on clayey or silty soils. Runoff should be controlled by prevention or collection where possible. However, runoff prevention by itself does not ensure infiltration and storage for use by crops ("green" water – Falkenmark, 1995) because some of the water temporarily stored on the surface may evaporate before it can infiltrate ("white" water – Falkenmark, 1995). In other cases, infiltrated water may move below the root-zone.

Several technologies and strategies have been developed that clearly demonstrate that the limited precipitation in dryland areas can be used more efficiently. However, they have not been widely accepted for various reasons. Unlike irrigation that generally results in large,

consistent, and predictable yield increases every year, dryland technologies may not result in any increase for one or even several succeeding years so farmers often become reluctant to continue the practice. An even more serious constraint for many dryland regions is the competing uses for crop residues. Many of the most successful *in situ* water conservation practices depend on leaving crop residues on the soil surface as a mulch to conserve water and enhance soil organic matter conditions. In many dryland areas, crop residues are critically needed for fuel or livestock feed and farmers perceive that these short-term benefits are greater than the long-term benefits that might result from sustaining the soil quality.

Kerr (2002) reviewed many watershed programs in India where water conservation technologies have been promoted vigorously. He stated that although the historic focus of most Indian soil and water conservation projects had been on mechanical measures such as trapping runoff water behind mechanical or vegetative barriers, it was widely recognized that conservation begins with sound agronomic practices such as maintaining soil cover and cultivating across the slope to encourage infiltration and reduce evaporation. However, Kerr (2002) surveyed farmers about a variety of conservation-oriented agronomic practices, including strict contour cultivation, cultivation across the slope, retaining stubble in the plot, and applying mulches to cover bare soil. Of all these practices, Kerr found that the only one practiced by more than a handful of farmers was cultivation across the slope. Farmers indicated that they recognized the value of applying mulches and retaining stubble in the fields throughout the dry season, but they rarely carried out these practices because of the high opportunity cost of forgoing use of the cut stubble for fuel and feed.

Even though *in situ* water conservation practices have not been widely accepted in many dryland regions because of various constraints, it is critically important that technologies and strategies for these areas continue to be developed and made known to farmers and policy makers so that they will become used when conditions warrant their adoption. In irrigated areas, transferring technology was relatively simple because improved seeds, fertilizers, and other inputs were quickly adopted because yields and profits were increased with little risk. In dryland areas, the success of technical interventions often depends on location-specific biophysical and socioeconomic conditions and often requires collective action by local people. Technologies that are successful one year may or may not be successful the succeeding year because of widely variable climatic conditions. However, the technologies presented below have proven effective in various dryland regions sufficiently often to warrant careful consideration.

TERRACES

Terraces (Plates 7 and 8) have been used for centuries as a way of controlling runoff and erosion. Because of the diversity in conditions where terraces are used, careful design is necessary to determine the most appropriate type of terrace for a specific location.

Bench terraces are perhaps the oldest type of terrace. They were used primarily in areas where the supply of agricultural land was limited and where population pressure forced cultivation up steep slopes. Early bench terraces were constructed by carrying soil from the uphill side of a strip to the lower side so that a level step or bench was formed. The steep slopes below the terraces were stabilized by vegetation or by neatly fitted stonework. Some early bench terraces are still being used successfully, e.g. radiocarbon dating indicates that the bench terraces in the Colca Valley in Peru were built at least 1 500 years ago (Sandor and Eash, 1995). The construction of terraces has continued in recent years, particularly in countries with limited land and high population pressure. In

China, more than 2.7 million ha of cropland were terraced from about 1950 to the end of 1984. This practice, combined with other measures of improved technology, resulted in a 2.8-fold increase in grain production (Huanghe River Conservancy Commission, 1988).

Despite their many benefits, the use of terraces has decreased in recent years for several reasons. They are costly to construct and maintain, furthermore terraced land is more difficult to farm, particularly with large equipment. The construction of terraces may also result in soil-fertility problems because topsoil is buried or moved downslope. Terraces are also subject to failure during large, intensive rainfall events, resulting in considerable damage that is costly to repair. Notable exceptions exist to the trend of not maintaining terraces, for example the level bench terrace system of the Colca Valley in Peruvian Andes (100 km²), and Zhuanland County, Gansu Province in China's loess plateau (1 555 km²) which have recently been rehabilitated (WOCAT, 2007).

In Yemen, one of the most extensively terraced areas in the world, there is a well-documented tradition of both dryland and irrigated farming over the past three millennia and much of the indigenous agricultural knowledge survives. Development efforts during the seventies and eighties in the north of Yemen focused on expansion of tubewell irrigation at the expense of the major land use on dryland terraces and traditional subsistence crops. Despite millions of dollars in aid, Yemen is far from agriculturally self-sufficient and its scarce water resource is

PLATE 7	PLATE 8
Terracing on an arid hillside	**The construction of terraces assists in soil and water conservation in the Syrian Arab Republic (M.Marzot)**

rapidly being depleted. Varisco (1991) explored the relevance of indigenous Yemeni knowledge of agriculture and the environment for the future of terrace farming in the country, arguing that farmer knowledge can contribute to sustainable production when integrated with modern methods and technologies. Within Yemen the existing community support networks and pride in national heritage would assist in a reinvestment effort for the existing resource of the terraces.

CONSERVATION BENCH TERRACES

Conservation bench terraces (CBTs) or Zingg terraces are a type of rainfall multiplier. They use a part of the land surface as a catchment to provide additional runoff onto level terraces on which crops are grown. The method is particularly appropriate for large-scale mechanized farming such as the wheat/sorghum farmlands of the southwest of the United States of America, where the method was pioneered by Zingg in 1955. Extensive trials in six western states with lowest rainfall compared the Zingg terraces with conventional level terraces and all-over bench terracing. The conclusion was that these types were as effective at controlling erosion as the other two practices, and more effective at reducing overall runoff. The data from these trials provided general guidelines on the method, but standard designs should be avoided because of the wide variation in the conditions of the soil, rainfall and farming system. The best way of applying the system in a particular situation should always be investigated locally.

The main application of the system is to increase the yield and the reliability of yield where rainfall is nearly sufficient for crop production (300–600 mm). Because of the high cost of installation of CBTs, it is not appropriate at very low rainfalls. Improving the probability of obtaining a reasonable crop may be more important than numerical increase in yield (FAO, 1987).

CONTOUR FURROWS

Contour furrows (or contour bunds and desert strip farming) are variations on the theme of surface manipulation that require less soil movement than conservation bench terraces, and are more likely to be used by small farmers, or in lower rainfall areas (Plate 9). The cropping is usually intermittent on strips or in rows, with the catchment area left fallow (FAO, 1987). The principle is the same as with CBTs, that is, to collect runoff from the catchment to improve soil moisture on the cropped area.

Where the contour furrows are not laid out precisely on the contour, or are built with some irregularities, there may be a danger of uneven depths of ponding behind the bank. This can be reduced by smaller bunds at right angles. However, as with tied ridging, these bunds should be lower in height than the main ridges so that any overtopping it will be laterally along the contour and not over the bund and down the slope. Sometimes, the emphasis is on the excavated furrow which collects water, so that in exceptional storms the runoff can overflow without damage.

CONTOUR BUNDS

Contour bunds (Plate 10) have been used in Kenya. At one site, a satisfactory sorghum crop was grown on only 270 mm of rainfall with a catchment ratio of 2:1. It was estimated that runoff from the catchment was 30 percent, giving 166 mm of runon, and 432 mm available to the plants (Smith and Critchley, 1983).

Contour bunds are also used in Ethiopia for a combination of soil conservation and water conservation. The bunds are built on a level grade with ties in the basin. A stone wall is built on the lower side of the earth bund in an attempt to reduce damage if the basin is overtopped (Hurni, 1984).

PLATE 9
Cabbage crop planted near contour furrows (L.Dematteis)

PLATE 10
Farmers working on the construction of contour bunds (G. Bizzari)

PLATE 11
Laser-levelled basin irrigation in the Tadla region, Morrocco (H.Bartali)

LAND LEVELLING WITH LASER AND MINI BENCHES

Land levelling with laser (Plate 11) is one of the most effective means of conserving runoff and preventing soil erosion (Box 1). However, because it is also the most expensive, this method has not been used widely except in areas with extreme land and water shortages. As a practical alternative to land levelling, narrow minibenches can be constructed economically on gentle (up to 2 percent) slopes (Jones, Unger and Fryrear, 1985). Soil cuts are relatively shallow and this reduces significantly the soil-fertility problems that are normally associated with the redistribution of large volumes of soil. Minibenches do not require much soil to be moved, making the system much less expensive to construct.

TIED RIDGES

Another alternative to land levelling is the use of furrow dyking, also called tied-ridging (Box 2). This is a proven soil- and water-conservation method under both mechanized and labour-intensive systems and it is used in many areas of the world. Furrow dykes retain precipitation on the soil surface until it can infiltrate. They are most effective where they are constructed on the contour. Seeding crops on the contour can be adapted to all types of tillage, including reduced-tillage and no-tillage systems, and this is highly recommended. Under mechanized systems, the furrow dykes are usually destroyed by tillage and have to be reconstructed each year. They can also become an obstacle during cultivation or harvesting operations. Perhaps the most important reason why more farmers do not adopt this technology is that, while the

BOX 1
Land levelling with laser in Morocco

New strategies for improving the already available irrigation systems are being devised in Morocco. For example, in order to improve on-farm irrigation, laser-levelled basin irrigation has been introduced on a number of farms in the Tadla region.

Laser-levelled basin irrigation has been widely used for field crops in the United States of America and for rice cultivation worldwide. It is particularly well adapted to flat terrain and heavy soils. Demonstrations on some farms showed substantial benefits in water saving of 20 percent and crop increases of 30 percent. Other farm inputs were improved by 10 percent and there were labour savings of 50 percent. The uniformity of irrigation was about 90 percent.

In principle, basin irrigation is the simplest of all surface irrigation methods. The key is to design the size of the basins to flood the entire area in a reasonable time, so that the depth of water is applied with a high degree of uniformity over the entire basin. Therefore, optimal sizes vary with soil types and stream flows. Very large basins served by flows of up to 150 litres/second are used in the United States of America. The method is not appropriate for crops that are sensitive to wet soil conditions around the stems or for crops on soils that crust badly when flooded. A disadvantage of basin irrigation is the interference of levees with the movement of cultivation and harvesting equipment (IPTRID, 2001).

additional effort is considerable, this does not increase yields in some years. Data from the Texas High Plains, USA, showed that the average runoff during the grain sorghum season was 25 mm; theoretically enough to increase grain yields by about 375 kg/ha (Stewart and Steiner, 1990). During the period of the study, there was little or no runoff in half the years and, often, two or three years without runoff occurred in sequence. Farmers often discontinue the practice before a sufficiently favourable response is obtained that would convince them to use the practice every year.

While the emphasis in semi-arid regions is usually on preventing runoff to increase the amount of water available for crop production, the prevention of runoff can lead to serious erosion problems if too much water accumulates. Water-management strategies must be site-specific. The most important factors are the soil-storage characteristics and the distribution of rainfall with respect to the growing season. In Hyderabad, India, El-Swaify et al. (1985) showed that the traditional cropping system on Vertisols resulted in 28 percent of the annual precipitation being lost as runoff, and 9 percent lost as deep percolation. On Luvisols (Alfisols), they found that the extent of runoff was similar (26 percent) but that percolation was 33 percent. There were substantial losses to percolation on both soils, and this occurred for all years of the study. The reason for these losses is that precipitation during the wet season exceeds the waterholding capacity of the soil profile. Eliminating runoff in these areas can result in serious waterlogging, particularly on Vertisols and other soils with high clay content. This is in contrast to other semi-arid regions where precipitation is often insufficient to fully recharge the soil profile.

Selecting the strategy for water conservation requires careful consideration of local conditions. Dhruva and Babu (1985) propose doing it by comparing rainfall with crop requirements giving three conditions:

- Where precipitation is less than crop requirements, the strategy includes land treatment to increase runoff onto cropped areas, fallowing for water conservation, and the use of drought-tolerant crops with suitable management practices.

- Where precipitation is equal to crop requirements, the strategy is local conservation of precipitation, maximizing storage within the soil profile, and storage of excess runoff for subsequent use.
- Where precipitation is in excess of crop requirements, the strategy is to reduce rainfall erosion, to drain surplus runoff and store it for subsequent use.

However, the weakness of this approach is that the main feature of rainfall in semi-arid regions is that it is very erratic and completely unpredictable (Brooks and Tayaa, 2002). There can be wide variations of moisture shortage and surplus both within and between seasons. A drought year whose total rain is well below the long-term average may still include periods of excessive rain and flooding, while a high-rainfall season may include periods of drought.

This makes the choice of strategy difficult, because the desired objective may change from

---

BOX 2
Experiences with tied-ridging

Most experience with the use of tied ridging has resulted in improvements. For example, in Africa, the system has been beneficial not only for reducing runoff and soil loss, but also for increasing crop yield (El-Swaify et al, 1985). However, in high rainfall years or in years when relatively long periods within the rainy season were very wet, significantly lower yields were reported from systems with tied ridges than from graded systems which avoided surface ponding of water. Under such conditions, tied ridging enhanced waterlogging, developed anaerobic conditions in the rootzone, resulted in excessive fertilizer leaching, and caused water-table rise in lower slope areas.

Macartney et al. (1971) reported that tied ridging in the United Republic of Tanzania gave higher maize yields in both low and high rainfall years. However, reports of success are more common in low rainfall years. For example, Njihia (1979) reported from Katumani in Kenya that tied ridging resulted in the production of a crop of maize in low-rainfall years when flat-planted crops gave no yield.

one season to another. In a dry area, it may be sensible to increase surface storage to improve crop yield in most years. However, in a wet year, this could cause waterlogging and reduce the yield. On the other hand, a drainage system may have the objective of increasing the runoff but also the undesired effect of exaggerating the effect of a drought. Therefore, it is not practical to classify methods according to average conditions, or to design strategies based on averages. Water management should reduce the problems caused by non-average events of flood and drought.

It may sometimes be possible to have dual -purpose strategies including methods that can be changed mid-season, for example, by opening up the ends of contour bunds to shed surplus water after a wet start to the season, or to block outlets for the opposite effect. However, not many methods allow this flexibility.

In addition to the variation in rainfall, there are other factors to consider: the soil, the land use, the farming system, and the social patterns (local lifestyles, social systems, and patterns of administration). Transferring what appear to be simple techniques requires not only the dissemination of information but also adaptation to local conditions.

WATER HARVESTING

Harvesting rainwater can be traced back to the 9th and 10th Century (GRDC, 2008). People in south and southeast Asia collected rainwater from roofs and from simple dams constructed from brush. Rainwater has long been used in the Loess Plateau regions in China where more recently, between 1970 and 1974, about 40 000 well storage tanks of various forms were constructed (GRDC, 2008). A thin clay layer was generally laid on the bottom of the ponds to minimize seepage losses and trees were planted at the edges of the ponds to help minimize evaporation (UNEP, 1982).

Perrier and Salkini (1991) defined water harvesting as a water-management technique for growing crops in arid and semi-arid areas where rainfall is inadequate for rainfed production and irrigation water is lacking. Rainfall is collected from a modified or treated area to maximize runoff for use on a specific site such as a cultivated field,

or for storage in a cistern or a reservoir, or for aquifer recharge. This definition is very restrictive and water harvesting is generally considered much more broadly. Bamatraf (1991) stated that farmers in Yemen tend to use water-harvesting techniques where rainfall is not sufficient. Thus, several approaches can be considered, including: runoff agriculture, where runoff is concentrated on a smaller area, generally used for arable or perennial crops; and runoff storage, generally in small reservoirs, used to supplement rainfall – often in horticulture or for livestock or domestic use. In dry farming, precipitation is captured in the soil where it falls during a fallow period and used to supplement rainfall during the next cropping period.

Water harvesting is sometimes practised with the primary objective to raise the water table to promote or sustain irrigation development. This has been the focus particularly in India and has been highly promoted and subsidized by many government and non-government programs. Although there have been some notable successes that have been widely publicized, the overall impact seems to have been minimal. Kerr (2002) and Batchelor et al. (2002) reviewed and summarized the impact of many watershed projects in India. Batchelor et al. (2002) acknowledged that different forms of water harvesting have been used successfully in semi-arid areas of India for millennia as a means of protecting domestic water supplies and increasing or stabilising agricultural production. Accepted wisdom has been that rainfall should be as far as possible be harvested where it falls and that these technologies are totally benign. They found, however, emerging evidence that water harvesting in semi-arid areas, if used inappropriately, can lead to inequitable access to water resources and, in the extreme, to unreliable drinking water. Kerr (2002) concluded that quantitative analysis did not yield strong conclusions about the success of water harvesting to develop irrigation. Kerr (2002) stated that none of the projects seem to have done much to assist farmers without irrigation or to help landless people gain access to the additional water generated through project efforts.

Water harvesting discussions in this publication will focus on harvesting water in surface structures for storage and subsequent use for growing crops or vegetables. The earliest water-harvesting structures are believed to have been built 9 000 years ago in the Edom Mountains in southern Jordan to supply drinking-water for people and animals (Oweis, 1996; Nasr, 1999). In southern Tunisia, ancient techniques such as meskat, micro-catchments, and jessour (terraces behind cross-dams in ephemeral watercourses) are still supporting olive and fig trees (Prinz and Wolfer, 1998). In Algeria, lacs collinaires (runoff storage ponds) have been used. In the caag system in the United Republic of Tanzania, floodwater from a stream is diverted and conveyed to a sequence of bunded basins used for cropping (Hatibu and Mahoo, 2000). The ancient hafirs (catchment reservoirs) in the Sudan (UNEP 2000) are still in use for domestic and livestock purposes as well as for the production of pasture and other crops. Siadat (1991) reported that in some parts of the Islamic Republic of Iran, such as Baluchistan in the southeast and Khorasan in the east, farmers have been using canal and dyke systems for centuries to spread water over parcels of cropland in order to increase soil-water storage. However, these techniques are practised in limited areas.

Water conservation and runoff storage were practised for centuries in India in an ecologically sound manner (Singh, 1995). The systems were decentralized, and the urban and rural communities played an active role in water management. Precipitation was the main source of water, most of it falling in a mere 100 hours in a year (Centre for Science and Environment, 2001). Once captured, this water met the demands for the rest of the year. These traditional water-harvesting systems declined when the provision of water with traditional decentralized systems was replaced with centralized systems. Centralized systems resulted in increasing and unsustainable dependence on groundwater sources and a gross neglect of the primary source of water – precipitation. An example is near Alwar in Rajasthan State. When the decision to sell off the trees was taken, the hills started to erode and could no longer hold the water during the few months of rains. The rivers stopped running and the wells went dry. When the wells went dry, the people who had depended on agriculture for many years could no longer grow food. In 1985, in an effort to reverse the process, the villagers began building johads, small dam-like structures. By 1986, the results were already visible. The rains

> **BOX 3**
> **Water harvesting schemes in the desert and semi-desert areas of Africa**
>
> This zone is characterized by very low, erratic rainfall (less than 500 mm/year). The major agricultural activity is nomadic stock raising (camels, goats, sheep and cattle) In this zone, the most important form of smallholder irrigation development is likely to be water harvesting. Spate irrigation could also be practised in occasional watercourses as well as shallow groundwater development. In some places, such practices are traditional. For example, in the Lower Omo Valley in Ethiopia (rainfall 300 mm/year), fodder and food-crop production depends almost entirely on seasonal floodwater from the River Omo and recession farming in the old river channels. Successful systems of runoff farming have also evolved in the adjacent Woito Valley.
>
> Water harvesting is often the only alternative to making a living through semi-nomadic stock-raising or depending on famine relief. An example of the establishment of water harvesting by semi-nomadic tribes was in Turkana in northern Kenya in the early 1980s. There were reports of local resistance in the form of apathetic attitudes and inherent scepticism towards the idea that crops could be produced on what was traditionally regarded as grazing land. Once there was demonstrable success in increasing fodder production, the level of participation grew rapidly and a large number of fields came under regular cultivation producing grain, fodder and wood (Oweis, Hachum and Kijne, 1999).

filled the johads, and the riverbed retained water for a much longer period. Within just a few years, the region once labelled a "black zone" by the Rajasthan government (meaning too dry to grow anything), again had a stable groundwater level, the five rivers in the region were again flowing continuously, and the villagers had returned to growing crops in the area.

In recent years, efforts have been made to implement modern techniques of water harvesting (Boxes 3 and 4). Al Ghariani (1995) reported very promising prospects for

BOX 4
Assessing the feasibility of water-harvesting techniques in Tunisia

In the arid regions of Tunisia, considerable investments are being made in maintaining the old water-harvesting techniques and introducing new ones to capture the scarce amount of rainwater (100–230 mm/year) for agricultural, domestic and environmental purposes.

A large variety of traditional methods (jessour, cisterns, etc.) and contemporary techniques (gabion, tabias, recharge wells, etc.) are found in the area. They have been playing various roles with regard to the mobilization and exploitation of rainfall and runoff waters (soil water, vegetation, flooding, aquifer recharge, etc.).

In most of the cases, the local population is aware of the environmental impacts of the introduction of new water-harvesting techniques. However, the real perception depends largely on the activity of the farmers (rainfed farming, irrigation, livestock, etc.) and their location (upstream, piedmont, downstream or coast) in the watershed. However, further refinements are needed to better include all possible impacts (positive and negative) that would occur as a result of the installation of the these structures. The interactions between upstream and downstream areas have to be addressed thoroughly to ensure partitioning of natural resources between different end users. The use of a geographical information system (GIS) and information technology would be of great value in transforming the obtained results as tool for decision-makers. The latter will be used in order to assess under which agro-ecological and socio-economic conditions investment in water-harvesting measures could be a viable undertaking in dry areas.

Source: Ouessar, Sghaier and Fetoui, 2002.

Jabal Al-Akhdar zone for cultivation of apple and cherry trees. The traditional stone walls and small collection basins have been improved and expanded. The main constraint on further development is the availability of skilled farmers to occupy and manage these newly established runoff-based farms.

In general, four types of water-harvesting techniques are used: micro-catchments, macro-catchments, floodwater harvesting, and rooftop water harvesting. These types of water harvesting are discussed briefly here, but details on the construction and application of these systems can be found in the FAO manual for the design and construction of water harvesting schemes for plant production (FAO, 1991).

MICRO-CATCHMENTS

Micro-catchment water-harvesting systems consist of a distinct catchment area and a cultivated area that are adjacent to each other (Hatibu and Mahoo, 1999) with the catchment being generally less than 1 000 m². The distance between the catchment area and the runoff receiving area is less than 100 m. These types of systems are simple, inexpensive and easily reproducible. Suleman *et al.* (1995) suggest that these systems offer significant increased cropping potential to smallholders without access to tractors in developing countries.

Several forms of micro-catchments (Box 5) have been used around the world: natural depressions, contour bunds, inter-row water harvesting, semi-circular and triangular bunds, meskats and negarims. The use of these will depend on the local conditions and the type of crop that receives the runoff water.

MACRO-CATCHMENT

Also called external catchments, macro-catchments (Plate 12) collect runoff from a large area located a significant distance from the cultivated area (Hatibu and Mahoo, 1999). The collected water is sometimes stored in a separate location before being used. Some types of external catchments include hillside-sheet or rill-runoff utilization, and hillside-conduit systems (Rosegrant *et al.*, 2002).

runoff agriculture systems in the Libyan Arab Jamahiriya. Successful trials have resulted in the construction of 53 000 ha of terraces around Tarhuna, Misallata, Urban and Assabas. Another 1 500 ha have been terraced in the

FLOODWATER HARVESTING

Floodwater harvesting within a streambed involves blocking the water flow, causing water to concentrate in the streambed (Plate 13). The streambed area where the water collects is then cultivated. It is important to make sure that the streambed area is flat with runoff-producing slopes on the adjacent hillsides, and that the flood and growing seasons do not coincide (Reij, Mulder and Begemann, 1988).

Ephemeral stream diversion is another external catchment system that is often used to harvest rainwater. In this technique, the water in an ephemeral stream is diverted and applied to the cropped area using a series of weirs, channels, dams and bunds (Box 6).

ROOFTOP WATER HARVESTING

- Rooftop water harvesting (Plate 14) is mainly used for domestic purposes and growing small vegetable gardens (Box 7). It is one of the most important options for addressing household food security in drought-affected, moisture-stressed environments. This is because:
- rainwater can be more easily available in moisture-stressed areas;
- the water captured requires low levels of external energy for extraction and transportation;
- the system can be easily implemented with family labour and using local materials;
- the system has low initial investment costs;
- the water can be used for other purposes.

FACTORS AFFECTING RUNOFF

Radder, Belgaumi and Itnal (1995) discussed the various factors governing the amount of runoff from a water-harvesting catchment area (Table 4) and summarized information from India on water-harvesting efficiencies of different surface treatments for inducing runoff (Table 5). The

---

BOX 5
Some examples of water harvesting
using micro-catchments

A decade of work in Jordan and the Syrian Arab Republic has demonstrated that micro-catchment techniques such as contour ridges for fodder shrub and pasture production have considerable potential for revegetation and combating degradation in rangelands. Where rainfall is less than 150 mm, micro-catchments economically support almond, pistachio and olive trees without supplemental irrigation. In the same area, water harvested and stored in small earth dams was used for the seasonal production of field crops. Rainfall-use efficiency can be very high using properly designed and managed water-harvesting systems. Overall system efficiency for small-basin micro-catchments in Jordan exceeded 86 percent. Where the system is not well designed and not managed properly, the efficiency drops to about 7 percent. These results reinforce the importance of combining technology development with the perceptions, needs and capabilities of the land users who will implement water harvesting (Oweis, 1997).

In the Province of Hamadan, the Islamic Republic of Iran, the use of runoff water (seilaub) is common. Rainwater is collected from sloping surfaces into channels running along slope-breaks and distributed to parcels located below the slope-breaks. In some places, water is stored in roughly constructed pools (estakhr) with a hole at the bottom that opens into a channel (djoob). The whole system is kept closed with a piece of wooden beam (dirak) which is pulled off to start irrigation. Similar pools are constructed at the openings of qanats with a low discharge capacity. As the water flows continuously, the pool remains full and can be used with a higher pressure where needed. Some of these techniques were used in the days of the ancient Persian Empire and are the product of local people's ability to manage scarce water resources on a sustainable basis (Farshad and Zinck, 1998).

In 1979, a small experimental area of micro-catchments for fuelwood trees was established with farmer participation in Burkina Faso. The farmers involved subsequently adopted these runoff-farming techniques and used them to improve their traditional erosion-control methods, thereby increasing their normal agricultural production. Fields long abandoned are now being reclaimed and farmers are increasing infiltration through the construction of simple contour bunds (Oweis, Hachum and Kijne, 1999).

PLATE 12
Upstream works enable subsequent action in
the valleys. This newly recovered seasonal pond
irrigated 80 ha in the 1993 rainy season
(F. Paladini and R. Carucci)

PLATE 13
A farmer clears a canal on the Nile River
(R. Faidutti)

most practical treatment was to compact the
soil surface; this resulted in a runoff coefficient
(the proportion of the precipitation leaving the
catchment area as runoff) of 30–60 percent.
The highest runoff achieved was about 90–95
percent when the soil surface was covered with
asphalt or fibreglass sheets. These treatments are
costly and require considerable maintenance.

BOX 6
Water spreading in eastern Sudan

In the Red Sea Province of Eastern Sudan,
traditional water-spreading schemes are
being rehabilitated and improved under the
technical guidance of the Soil Conservation
Administration. In this semi-desert region,
wadi beds are the traditional sites for cropping
of sorghum by the semi-nomadic population,
using the moisture from natural flooding.

The improvements being introduced are
large earth diversion embankments in the
main or subsidiary channels, and then a series
of spreading bunds or "terraces". These bunds
are usually sited on, or approximately on, the
contour, and spread the diverted flow. The
spacing between bunds in the almost flat
landscape is not fixed but can range up to
200 m apart. Bunds are usually up to 150 m
long and at least 75 cm high. Some machinery
is employed, but manual labour supported by
incentives is also used (FAO, 1991).

More recently, Oweis, Prinz and Hachum (2001)
estimated runoff coefficients and costs of typical
runoff-inducement techniques using information
from locations in the Near East (Table 6). These
coefficients provide information necessary for
evaluating the potential of harvesting water for
a given region, and the results can be coupled
with water-use efficiency values to estimate
the production from the harvested water. For
example, if the runoff coefficient of an inducement
technique is 50 percent, and it is estimated that
200 mm of annual precipitation is subject to
water harvesting, then 1 000 m³/ha of water
could be harvested. The water-use efficiency of
producing wheat grain varies considerably but a
reasonable estimate is about 1.3 kg/m³ (Musick
and Porter, 1990). Therefore, about 1.3 tonnes of
wheat grain could be produced from the water
harvested from a hectare of runoff area.

Although land treatment has a major impact on
the runoff coefficient, the size of the contributing
area and the intensity of the rainfall event are
also major factors in determining the amount of
runoff. A study in the Negev Desert (Ben-Asher,
1988) reported a runoff coefficient of 70 percent
where the contributing area was 100 m², but
only about 15 percent where the contributing
area was 10 000 m². FAO (1991) showed a
relationship between the amount of rainfall for a
particular event and the runoff coefficient. That
work showed a runoff coefficient of 35 percent
when 50 mm of precipitation was received, but
only 5 percent when 15 mm of precipitation
fell. These values will change depending on the

surface treatment. Where the contributing area is covered with plastic or other material that is impermeable, the differences will become much smaller and in some cases the runoff coefficients can approach 95 percent.

REDUCING EVAPORATION

Evaporation is a major cause of water loss in semi-arid regions. The goal of efficient water use in semi-arid regions should always be to maximize the percentage of annual precipitation used for transpiration by decreasing losses from runoff, evaporation and percolation (El-Swaify *et al.*, 1985). In most semi-arid locations, evaporation from the soil is the largest loss (Figure 8.) In addition to water loss by evaporation during fallow periods, there are significant losses during the crop-growing period. Water loss by soil evaporation during the growing season is highly dependent on the leaf area index (LAI). The LAI is the total area of green leaves per unit area of ground covered, usually expressed as a ratio (WMO, 1990). At a LAI of less than two, half or more than half of the evapotranspiration is evaporation from the soil surface (Ritchie, 1983). Evaporation from the soil surface can be as much as 20 percent even at a LAI of 3 or more.

For sorghum, Hanks, Allen and Gardner (1971) found about 36 percent evaporation from the soil surface at a leaf area index of 1.2. Using a computer simulation model, Stewart and Steiner (1990) estimated that 30–35 percent of the evapotranspiration for grain sorghum grown at a high soil-water level was lost as evaporation,

BOX 7
Rooftop water collection for food security

A South African case
Laying a thin cement surface around rural homes is effective in capturing rainwater and can feed into underground storage tanks. Roof water can be captured in the same manner, eliminating the need for gutters. With an annual rainfall of 500 mm, impermeable surfaces of 100 m² – approximately the area of the roof and lapa (paved area) of a modest-sized rural home – can yield 50 m³ of water during South Africa's dry season. This is sufficient to irrigate a vegetable garden and contribute towards food security for poorer families. This method has the added benefit of providing relatively clean and sediment-free water (IWMI, 2003).

Water-storage structures
Tanks may store water collected from ground surfaces, tin rooftops, greenhouses, springs and rivers. The stored water can be used for irrigating crops (supplementary, full irrigation or both), supplying water for livestock and household needs or any combination of these.

Depending on their size and type, water tanks may serve individual households, groups of them, schools, hospitals or the whole community. In general, larger tanks cost more than individual structures, but are cheaper per cubic metre of water stored. They are also more difficult to construct and manage. Although the use of plastic bags has proved useful in India, a balance between economy and durability should be considered when designing storage tanks.

An example of a cost-effective water-storage structure is the externally reinforced brick tank developed by the University of Warwick (United Kingdom). Supported by a packaging strap, this structure is able to withstand internal stresses. As a result, it requires less material in construction. Modern construction and design can also improve indigenous methods of water storage. For example, hand-dug wells can be lined with steel barrels, cement bricks or steel-reinforced concrete for greater durability (IWMI, 2003).

TABLE 4
Factors governing the amount of runoff from a water-harvesting catchment

Major factor	Associated factors
1. Rainfall	Rainfall intensity Rainfall duration Rainfall distribution Events of rainfall causing runoff
2. Land topography	Degree of land slope Length of run Size and shape of the catchment Extent of depressions and undulations of the catchment
3. Soil type	Soil infiltration rate Antecedent soil moisture Soil texture Soil structure Soil erodibility characteristics
4. Land-use pattern	Cultivated, uncultivated or partially cultivated Under pasture or forests Bare, fallow or with vegetation Soil- and moisture-conservation measures adopted or not Crop cultural practices adopted

Source: Adapted from Radder, Belgaumi and Itnal, 1995.

and that 40–45 percent was lost as evaporation under intermediate soil-water conditions. A major strategy for increasing yields in dryland regions should be to reduce evaporation losses during both the fallow period and the crop-growing season.

The land in the Great Plains of the United States of America was tilled repeatedly during fallow periods to control weeds so that soil-water levels would be increased for the subsequent wheat crop. However, these practices left the soil bare and caused a rapid decline in SOM, triggering extensive wind erosion during the

TABLE 5
Water-harvesting efficiencies of surface treatments for enhancing runoff

Surface treatment of the catchment	Water harvesting efficiency[a] (%)
1. Compacted soil surface	30–60
2. Removing the vegetation	7–21
3. Roaded catchment	24–41
4. Coating of bitumen on compacted soil	65–89
5. Sodium chloride	10–64
6. Sodium carbonate	35–71
7. Mixture of bentonite clay and sodium chloride	48–61
8. Bentonite clay	19–56
9. Asphalt	60–90
10. Asphalt fibreglass sheet	85–95
11. Asphalt roofing	52
12. Bitumen with kerosene soil	77
13. Concrete membrane	56–80
14. Low-density polyethylene sheet	60–85
15. Silicane and paraffin	50–80

[a] Percent of precipitation falling on catchment area harvested. Source: Modified from Radder, Belgaumi and Itnal, 1995

drought years of the 1930s: the infamous Dust Bowl, a major human-exacerbated ecological disaster (Stewart, Jones and Unger, 1993). In order to combat this problem, stubble mulching became widespread: pulling flat V-shaped sweeps or blades through the soil about 10 cm beneath the surface. This operation cuts plant roots and kills the weeds but does not invert the soil. Therefore, much of the crop residue is left on the surface as a mulch to protect against wind and water erosion and reduce evaporation loss. Even relatively small quantities of residues are highly effective in the control of both wind and water erosion (Plate 15). As a guideline, plant residues covering 30 percent of the soil surface will reduce both wind and water erosion by about 80 percent (Laflen, Moldenhauer and Colvin, 1981; Fryrear, 1985). Although stubble mulching was developed to address the wind erosion problem, it soon became evident that the mulch increased soil-water storage as well. This is attributed to increased infiltration as well as to reduced evaporation. The contribution of each of these factors will vary with specific conditions

Mulches left on the soil surface – or dust mulch by repeated ploughing under certain conditions as in India (Annex 2) – have proved effective in reducing evaporation during fallow periods. Other studies have also shown that leaving crop residues on the soil surface reduces evaporation and increases soil-water storage (Cornish and Pratley, 1991; Li Shengxiu and Xiao Ling, 1992; Smika, 1976). Unger and Parker (1976) showed that wheat stubble was about twice as effective in decreasing soil-water evaporation as grain sorghum stubble and more than four times as effective as cotton stalks. The differences resulted primarily from the physical nature of

the residues (hollow, pithy or woody), which affected the specific gravity and, hence, their thickness and surface coverage when applied at identical rates by weight.

Surface residues are most beneficial for reducing evaporation when several precipitation events occur over a period of a few days. This allows each successive precipitation event to wet the soil to a greater depth. Water stored at greater depths is less vulnerable to evaporation and, therefore, more likely to be available during the next cropping season.

Achieving a reduction in evaporation during the growing season is somewhat more complex than during a fallow period. Unger and Jones (1981) evaluated the effect of straw mulch during the growing season on growth, yield, grain quality, water use and water-use efficiency of grain sorghum. Sorghum responded more to the amount of soil water at time of seeding than to the presence of mulch during the growing season. A positive impact of mulch was found mainly on the plots with a low water level. The authors concluded that shading from the plant canopy largely substituted for the beneficial effect of mulch during the growing season. Therefore, the best strategy for increasing the transpiration portion of evapotranspiration is to establish a plant canopy as quickly as feasible, e.g. through narrower row spacing and higher plant populations. However, these practices can lead to lower water-use efficiencies when the harvestable product is grain because the soil water may become depleted prior to grain filling. This dilemma is faced by dryland crop producers in selecting the proper row width and plant density.

TABLE 6
Estimated runoff coefficients and cost of runoff-inducement techniques

Treatment	Runoff coefficient	Estimated life	Cost
	(%)	(years)	(US$ per 100 m²)
Catchment clearing	20–35	1–3	1–4
Surface smoothing	25–40	2–4	2–4
Soil compaction	40–60	2–3	6–10
Surface modification	70–90	3–5	10–20
Surface sealing	60–80	5–10	4–10
Impermeable cover	95–100	10–20	20–100

Source: Adapted from Oweis, Prinz and Hachum, 2001

PLATE 15
Smallholder coffee farmers cover the ground with
straw to preserve humidity, Malawi (A. Conti)

A better practice, particularly for summer crops such as grain sorghum, may be to choose cultivars with a shorter growing period. These can be seeded at a higher plant population, resulting in a more complete canopy at an early stage. Studies in the semi-arid Texas High Plains in the United States of America by Jones and Johnson (1997) suggest such a short-season, high-density strategy for dryland grain–sorghum production. Short-season hybrids have a lower genetic yield potential than long-season hybrids, but they have a higher harvest index and use water over a shorter period. This places less reliance on individual growing-season precipitation and more reliance on stored soil water to produce grain. Jones and Johnson (1997) stated that the short-season, high-density strategy for successful grain sorghum production in the Texas High Plains requires a soil profile with 125–200 mm of stored plant-available soil water and an anticipated growing season precipitation of at least 250 mm.

INCREASING SOIL ORGANIC MATTER CONTENT AND FERTILITY

The primary repository of soil fertility is soil organic matter (SOM). A decrease in SOM is an indicator of declining soil quality because SOM is extremely important in all soil processes biological, physical, and chemical. It acts as a storehouse for nutrients, improves nutrient cycling, increases the cation-exchange capacity and reduces the effects of compaction. It builds soil structure increasing the infiltration and water storing capacity. It serves as a buffer against rapid changes in pH and an energy source for soil micro-organisms.

BOX 8
Growing sorghum plants in clumps
to increase grain yield

Recent studies by Bandaru et al. (2006) have shown that growing grain sorghum in clumps of 3 or 4 plants instead of equally spaced plants can increase grain yield significantly under certain semiarid conditions.

In the southern Great Plains of the United States, stored soil water and growing season precipitation generally support early season growth but are insufficient to prevent water stress during critical latter growth stages. Growing plants in clumps compared to uniformly spaced plants reduces the number of tillers and vegetative growth. This preserves soil water until reproductive and grain-filling stages, which increases grain yield. There are marked differences in plant architecture of uniformly spaced plants compared to clumped plants. Uniformly spaced plants produce more tillers and the leaves on both the main stalk and tillers grow outward, exposing essentially all the leaf area to sunlight and wind. In contrast, clumped plants grow upward with the leaves partially shading one another and reducing the effect of wind, thereby reducing water use. In the studies by Bandaru et al. (2006), grain yields were increased by clump planting by as much as 100 percent when yields were in the 1000 kg/ha range and 25 to 50 percent in the 2000 to 3000 kg/ha range, but there was no increase or even a small decrease at yields above 5000 kg/ha.

An annual loss by decomposition of 1–2 percent of the organic matter in the surface 15 cm of cultivated soils is not uncommon. In some climates, the loss can be considerably higher. For example, Pieri (1995) summarized data from semi-arid regions of Africa and reported that on highly sandy soils, annual ploughing with application of fertilizers led to an annual loss of 5 percent or more in organic matter. Only practices with manure applications prevented a decline in SOM. The effect of ploughing by itself was less clear, but several of the reported studies indicated that ploughing increased the rate of decline. Pieri (1995) proposed that there is a critical level for SOM dependent on the sum of the clay and silt contents. Where the SOM percentage falls below the critical level, the maintenance of soil structure becomes difficult. However, he disagreed with agronomists who argue that, as SOM is important in soil quality, the higher the SOM content, the better the soil. Pieri stated that it is fruitless to aim for a SOM percentage above the critical level in semi-arid Africa, where there are many other technical and economic constraints on crop performance. Most drylands soils have been depleted in SOM due to inappropriate cultivation, overgrazing and/or deforestation in the past, causing a decline in soil quality and emission of C into the atmosphere. There is great potential to increase the SOM of most dryland soils before such a critical level (equilibrium) is reached (Lal, 2002a).

One of the problems with crop production in drylands is determining whether there are favourable moisture interludes in the cycle of plant development when the soil cannot supply sufficient nutrients. Total yield and water-use efficiency can be increased where fertilizers can increase the net assimilation rate or growth in these periods without exhausting water at a faster rate. If fertilizers accelerate the rates of growth and water use, the yield and water-use efficiency will depend on the total supply of water and the status of the crop when the water supply becomes exhausted. Thus, accelerated water use through fertilization can be disastrous for grain crops if the soil water supply is exhausted and rainfall events do not occur in time for grain filling. This timing of water use, total water supply and plant development is much less critical for crops that are grown for their vegetative parts and do not need to complete their life cycle through seed production.

In studies of fertilizers and the efficient use of water, various authors in Kirkham (1999) concluded that any practice that increases dry matter production would lead to increased water-use efficiency. Exceptions are those cases where water greatly in excess of consumptive-use demands is essential in attaining that production, such as frequent irrigations after planting to establish small-seeded crops or leaching with irrigation water to remove soluble salts. The conclusion by Viets (1962) that increases in dry matter production increases water-use efficiency is also reflected in the relationships in Figure 7, which show that cereal yields can be significantly different when the same amounts of water are used. Differences in yields occur when inputs are added to remove other constraints. However, it is important to note that the added inputs do not have a marked effect unless the water constraint is addressed (Kirkham, 1999).

Long-term agricultural experiments in Europe and North America indicate that soil organic matter and carbon are lost during intensive cultivation. Losses typically show an exponential decline following the early years of cultivation of virgin soils, with continuing steady losses over many years (Arrouays and Pélissier, 1994; Reicosky et al., 1995); Reicosky, Dugs and Torbert, 1997; Rasmussen et al., 1998); Tilman, 1998; Smith, 1999; Pretty and Ball, 2001). It has also been established that SOM and soil carbon can be increased to new higher equilibria with sustainable management practices. A wide range of long-term comparative studies show that organic and sustainable systems improve soils through the accumulation of organic matter and soil carbon, with accompanying increases in microbial activity, in various locations: the United States of America (Lockeretz, Shearer and Kohl, 1981; Wander, Bidart and Aref, 1998; Petersen, Drinkwater and Wagoner, 2000); Germany (El Titi, 1999; Tebrügge, 2000); the United Kingdom (Smith et al., 1998; Tilman, 1998; Scandinavia (Kätterer and Andrén, 1999); Switzerland (FiBL, 2000); New Zealand (Reganold, Elliott and Unger, 1987; Reganold et al., 1993); only a small number of studies have been undertaken in the tropics (Chander et al., 1997; Post and Kwon, 2000).

The importance of soil organic matter is difficult to overemphasize, particularly in semi-arid regions and its maintenance is clearly a major constraint on the development of sustainable agro-ecosystems. Despite the many proven benefits of SOM, its management and recycling in an intensified, modern agro-ecosystem must necessarily revolve around two fundamental characteristics:
- the on-farm availability of organic material;
- and the economic incentive for conserving and recycling organic matter.

Perhaps the two most important practices for maintaining SOM are to minimize soil disturbance and to apply organic (animal, human and vegetal) wastes. Other practices that lead to increased yields are also important, because more carbon will be added to the soil from increased root production and crop residues (Plate 16). The high demand for crop residues in many developing countries for fuel and animal feed makes it particularly challenging to maintain SOM. This problem is likely to be further exacerbated in future as the new demand for residues to produce second generation biofuels develops. Where feasible, it is better to have animals graze crop residues *in situ* so that the manure is distributed over the area, rather than to remove the residues for feeding off-site. When it is necessary to remove the crop residues for use as livestock feed (i.e. in lot or zero grazing systems), every attempt should be made to return the manure produced to the land. Otherwise, the SOM content will continue to decline and may reach a level where the long-term sustainability of the soil-resource base becomes threatened.

Most indigenous soil- and water-conservation practices in drylands have tillage as their centrepiece. Since the early 1960s, however, scientists and farmers have been developing forms of conservation tillage to:
- reduce production costs and increase profit margins for farmers;
- reduce runoff and associated losses of soil, water, seeds, applied inputs and organic matter;
- reduce wind erosion and wind erosion air quality degradation;
- improve environment for root development, including better availability of plant nutrients in the root zone, better infiltration and water-holding capacity of soils, and reduced amplitude of day-to-night temperature ranges;
- increase efficiency of use of the available water;
- reduce the amount of fossil fuels used in growing food; and
- maintain or enhance soil organic matter.

Conservation tillage and conservation agriculture are often used as umbrella terms commonly given to no-tillage, minimum tillage and/or ridge tillage, to denote that the inclusive practices have a conservation goal of some nature (Baker *et al.*, 2007). Usually, the retention of at least 30 percent ground cover by residues after seeding characterizes the lower limit of classification for conservation tillage or conservation agriculture, but residue levels alone do not adequately describe all conservation tillage or conservation practices and benefits.

Conservation agriculture (CA) is specifically defined by FAO as a system that aims to achieve sustainable and profitable agriculture and subsequently aims at improved livelihoods of farmers through the application of the three CA principles: minimal soil disturbance, permanent soil cover and crop rotations (FAO, 2007). In reality, some conservation systems do not always employ all three of these principles. Studies have shown that successful implementation of these principles promotes infiltration of rainwater, reducing or eliminating runoff (blue water) and erosion, also reducing evaporation (white water) and lowering soil surface temperatures – conditions for more effective for soil and water conservation (Lal, 1997; FAO, 2004). With time under CA, soil life assumes the functions of human / mechanical soil tillage, loosening the soil and mixing the components. The adoption of conservation farming has generally been slow – attributed primarily to the fact that the routine of tillage in conventional agricultural systems, whether by hand, ox- or tractor drawn plough, is so heavily ingrained in the culture of arable farming communities.

PLATE 16
Smallholder farmers incorporate crop residues in the soil to improve soil fertility, Malawi (A. Conti).

Plate 17
As a farmer ploughs, serious erosion eats away the land.

C than vegetation and twice as much as that which is present in the atmosphere (Batjes and Sombroek, 1997). Soils contain 1 500Pg of C to 1m depth and 2 500Pg of C to 2m (1Pg = 1 gigatonne); vegetation contains 650Pg of C and the atmosphere 750Pg of C.

Conservation agriculture, zero and low tillage agricultural systems in all farming systems provide a sink for the growing atmospheric concentrations of carbon dioxide (CO_2) which are driving climate change (Lal, 1997; Schlesinger, 2000; FAO, 2004; Stern, 2006). This benefits land users directly, as they improve the organic matter status of their soils, improving fertility and water storing

Research from several countries shows significant improvements in crop yields and reduced soil erosion, also lowering peak labour demand and reducing labour requirements after the introduction of tied ridging or pitting to increase infiltration, followed by the adoption of zero- or minimum-tillage (direct-planting) systems (Kaumbutho and Simalenga, 1999). Conservation agriculture is a promising approach for redirecting the components of the water balance in favour of infiltration and consequently crop transpiration (green water) and production (WOCAT, 2007). Experience has shown that CA systems achieve yield levels as high as comparable conventional agricultural systems, but with less fluctuation due for example to drought, storms and floods. CA therefore contributes to food security and poverty reduction, reducing the risks for the communities (health, living conditions and water supply) and also the costs for the State (less need for road maintenance and emergency assistance).

The general population of the district, state or river basin also gain considerable benefits from positive externalities of widespread conservation agriculture (FAO, 2002). These include: less downstream sedimentation; more regular river flows; aquifer recharge; reduced air pollution; increased carbon sequestration; and conservation of terrestrial and soil-based biodiversity.

Soils are the largest carbon reservoir of the terrestrial carbon cycle. The quantity of C stored in soils is highly significant on the global scale; soils contain about three times more

> **BOX 9**
> **Effects of tillage on soil**
>
> Tillage results in a rapid decline in SOM, particularly in hot regions. Tiessen, Cuevas and Chacon (1994) reported that soil carbon contents in Canadian prairie soils had decreased by about 50 percent as a result of 65 years of cultivation. In contrast, only 6 years of cultivation in a Brazilian semi-arid thorn forest reduced the soil carbon content by 40 percent (Wood, Sebastian and Scherr, 2000). Nevertheless, the plough remains the symbol of agriculture, and tilling the soil has been hailed as the most effective way of controlling weeds and improving soil fertility. In the early years of cultivation, soil fertility may be adequate because the decomposition of SOM releases all the nutrients required for plant growth. However, these nutrients are nothing more than the debris of decomposed SOM. Unless the fallow period is long enough, the soil fertility declines rapidly in dryland regions following cultivation. At the same time, the soil physical properties deteriorate and make the already limited water less effective. The hazards of wind and water erosion are also increasing. Growing demographic pressure is causing persistent land degradation. As a result, farmers in many areas may be experiencing agricultural drought even when there is no meteorological drought, with crops suffering from a scarcity of plant-available soil water even when there is adequate precipitation.

capacity, reslting in more reliable crop yields. Often without being aware of it, CA practitioners are contributing to mitigating the effects of GHG emissions from the burning of fossil fuels. Lal (2004) calculated that an increase of 1 ton of soil carbon pool of degraded cropland soils may increase crop yield by 20 to 40 kilograms per hectare (kg/ha) for wheat, 10 to 20 kg/ha for maize, and 0.5 to 1 kg/ha for cowpeas. As well as enhancing food security, carbon sequestration has the potential to offset fossil fuel emissions by 0.4 to 1.2 gigatons of carbon per year, or 5 to 15 percent of the global fossil-fuel emissions.

In the case of drylands, the lack of water severely constrains plant productivity and affects the accumulation of C in dryland soils (FAO, 2004). Consequently dryland soils contain relatively small amounts of C (between less than 1 percent and less than 0.5 percent (Lal, 2002b). The organic matter content of dryland soils will rise with the addition of biomass to a soil which has previously been depleted due to land use change (e.g. conversion from natural vegetation to arable). Although the rate at which carbon is sequestered is low in drylands compared with soils of temperate regions, the potential offered by drylands to sequester C is large, not only because of the large geographical extent, but because historically, soils in drylands have lost significant amounts of C and are far below their critical level (FAO, 2004 and Oldeman et al. (1991).

Although conservation agriculture and other types of conservation systems offer substantial benefits, adoption has been slow. FAO (2001b) reported that conservation agriculture was being practised on about 45 million ha in 2000, or about 3 percent of the 1 500 million ha of arable land worldwide. The transformation from conventional tillage to conservation agriculture requires farmers to acquire considerable management skills and involves investment in new or modified equipment or tools. It also requires a higher level of management, and perhaps most important, a change in the mindset of farmers.

THE IMPORTANCE OF CROP AND CULTIVAR SELECTION

Water- and soil-management strategies should be accompanied by using appropriate crops and cultivars with optimal physiology, morphology and phenology to match local environmental conditions. Breeding and selection for improved water-use efficiency and the use of genotypes best adapted to specific conditions can improve soil-water use and increase water productivity (Studer and Erskine, 1999).

An important approach to increasing the efficiency

BOX 10
Early sowing of chickpea

In the Mediterranean region, rain falls predominantly in the cool winter months of November–March. Traditionally, chickpea is sown in late February and early March. As a consequence, the crop experiences increasingly strong radiation and a rapid rise in temperature from March onwards. This causes the rate of leaf area development to increase with consequent high evapotranspiration. This period of high evaporative demand occurs at the end of the rainfall when the residual soil moisture is inadequate to meet the evaporative demand. Therefore, the crop experiences drought stress during late vegetative growth and reproductive growth, resulting in low yields. The replacement of traditional spring sowing with winter sowing is possible but only with cultivars possessing cold tolerance and resistance to key fungal diseases [For chickpea, specifically breeding is for tolerance to Ascochyta Blight (Singh and Ocompo, 1997; Studer and Erskine, 1999).]

The average gains in seed yield from early sowing chickpea over three sites and ten seasons is 70 percent, or 690 kg/ha, which translates into an increase in water-use efficiency of 70 percent (Erskine and Malhotra, 1997). In 30 on-farm trials comparing winter with spring chickpea in north of the Syrian Arab Republic, the mean advantage of winter sowing in seed yield and water use efficiency was 31 percent (Pala and Mazid, 1992). Currently, an estimated 150 000 ha of chickpea is winter-sown in the West Asia and North Africa regions.

of water use is to change both water-management practices and cultivar concurrently. This allows a considerable increase in productivity. Seasonal shifting, i.e. the development of crop varieties that can be grown in winter under lower evaporative demand, represents an additional challenge for breeders seeking to use scarce water more efficiently as traits such as winter hardiness and disease resistance have to be improved. Early and complete canopy establishment to shade the soil and reduce evaporative loss from the soil surface can significantly improve the water productivity of rainfed crops in Mediterranean conditions and also that of summer-rainfall crops over much of the semi-arid tropics (Cooper *et al.*, 1987 and Oweis *et al.*, 2001).

An alternative approach, particularly appropriate for subsistence smallholders in drylands is to resume growing the wider range of more traditional grain crops and legumes, which are better adapted to dry land conditions, not restricting themselves to the small range of varieties of crops which have become ubiquitous in the late twentieth century (wheat, barley, sorghum, maize) and legumes (chickpea and clovers)

Agrobiodiversity is a vital subset of biodiversity (CBD, 2007 and FAO, 2005). Sources of stresses are numerous in the drylands (drought, insect attacks, diseases, high temperatures, off-season rain). One of the major ways farmers can minimize risk is by growing a diversity of crop species and varieties.

The implications of the rapid reduction in agrobiodiversity during the twentieth century (particularly post Green Revolution) have been profound, increasing the risk of harvest failures due to drought, disease and / or pests. Raising awareness of the importance of local agrobiodiversity will contribute to reducing the risk of crop failure in the coming decades.

ROLE OF INDIGENOUS AND INTRODUCED PRACTICES

Local farmers are the key individuals with responsibility for improving soil and water conservation and management in developing countries. These small-scale farmers are highly diverse. Even within small communities, individual

farmers have a wide range of circumstances, including their needs, priorities, availability of resources and also preferences. Farmers each have a wealth of knowledge about their crops, their soils, their farming environment, also diverse socio-economic conditions. They use this knowledge as the basis not only for making decisions and communicating with one another, but also in many cases as the basis for innovation. Small-scale farmers are keen observers and conduct experiments on their own (Reij and Waters-Bayer, 2001). Policy-makers and scientists must understand and appreciate the depth of local knowledge before they can communicate with the farmers to acquaint them with new or improved technologies. Large-scale top down approaches (transfer of technology models) to development have repeatedly been shown not to succeed, often as they are too costly for small-scale farmers to implement – or they do not take into account local factors (Reij and Waters-Bayer, 2001). Farmer field schools which encourage learning-by-doing are increasingly proving successful to help smallholders learn new information, particularly in the field of integrated pest management and conservation agriculture (Van de Fliert, 1993; Feder *et al.*, 2004); Simpson and Owens, 2002). Understanding and trust between all parties must be established before farmers can be expected to test, adapt and adopt new or improved technologies.

Indigenous practices refer to local practices (Plate 18), as distinct from interventions initiated

PLATE 18
A farmer clears a canal to ensure water flow through a falaj system in Oman. The falaj system in the Al-Jauf region dates back 2 500 years

from outside (Scoones, Reij and Toulmin, 1996). However, many practices regarded as indigenous today may have been derived from elsewhere in the past (Oweis *et al.*, 2004). They become indigenous once they have been adapted to fit local conditions, widely accepted by local farmers and used for many years. In essence, these practices become part of the local culture and are not easily changed. Introduced practices are often specified in technical manuals and extension handbooks with precise dimensions and design requirements. Indigenous practices are much more flexible. Flexibility is important, as field topography and other biophysical and socio-economic conditions vary from site to site.

For the development planner and project administrator, the use of an off-the-shelf technical package might be appealing. However, when new technologies are introduced, and particularly where they are imposed, problems arise. The reasons for problems vary widely from setting to setting. Scoones, Reij and Toulmin (1996) found that where land is in plentiful supply, or where the cultivator can easily move into other fields of economic activity, there may be little long-term interest in maintaining soil fertility. Areas with high population densities and few options outside agriculture often had elaborate soil- and water-conservation structures. In contrast, the level of labour investment for water harvesting and other practices was far lower in areas with low population density. In many cases, the tasks related to indigenous practices were divided

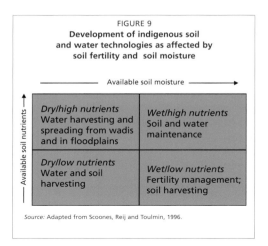

FIGURE 9
Development of indigenous soil and water technologies as affected by soil fertility and soil moisture

Source: Adapted from Scoones, Reij and Toulmin, 1996.

according to gender and introduced practices may interfere with this balance (Scoones *et al.* 1996) and IWMI, (2006).

Figure 9 summarizes the development of indigenous soil- and water-management practices. Where both soil moisture and soil fertility are low, indigenous practices focus on both soil management and water harvesting. Where soil moisture is low but fertility is high, the focus is on water harvesting. However, in both cases, there are few or no inputs other than labour. As soil moisture becomes less of a constraint, the management focuses more on fertility and soil and water maintenance. Scoones, Reij and Toulmin (1996) also compared the characteristics of indigenous and introduced soil- and water-

TABLE 7
Characteristics of indigenous and introduced conservation practices

Characteristics	Introduced practices	Indigenous practices
Designed by	Engineers and development planners	Local farmers
Designed for	Soil conservation	Multiple, depending on setting (including soil/water harvesting, conservation, disposal)
Design features	Standardized in relation to slope features	Flexible, adapted to local microvariation
Construction	One-time	Incrementally (fitting with household labour supply)
Labour demands	High	Variable, generally low
Returns	Long-term environmental investment	Immediate returns
Project setting	Large-scale, campaign approach; food-for-work / cash-for-work/ employment-based safety-net programmes, etc.	Longer-term support to indigenous innovation; participatory research and farmer-to-farmer sharing

Source: Adapted from Scoones, Reij and Toulmin, 1996.

conservation practices (Table 7), indicating that soil- and water-conservation practices often face serious constraints.

Kerr and Sanghi (1992) reviewed indigenous soil- and water-conservation practices in six regions of India's semi-arid tropics and found that these were generally preferred to introduced practices (Box 11). The indigenous practices evolved in different ways from place to place in response to local agro-ecological and economic conditions. Three factors were common among all locations:

- Firstly, the designs of indigenous practices reflected the relative availability and opportunity costs of different agroclimatic factors and resources, including materials, human labour, animal power and cash.
- Secondly, practices developed within the constraints of small, fragmented farms in accordance with farmers' preferences to invest in soil and water conservation individually or in cooperation with an adjacent farmer rather than in large cooperative groups.
- Thirdly, economic factors determined adoption patterns.

Investments in soil and water conservation and management are one among a range of economic concerns. Farmers assimilate available information in deciding how their time and money can be spent most productively. Their opportunities and constraints are not identical, so the same activity is not equally profitable for all farmers. Often, soil- and water-conservation practices introduced by outside groups or organizations have a single objective. In order to meet this objective, technologies are introduced with designs to conserve the maximum amount of soil and water. In contrast, farmers have multiple objectives that may include soil and water conservation.

One indigenous practice relating to soil and water conservation, water harvesting and water management is that of using "qanats". This is an irrigation system that was developed in Persia some 2 000 years ago and then spread to central Asia, China and North Africa. This indigenous practice takes advantage of the rainfall and groundwater resources in arid regions bringing water resources to the surface by gravity through carefully designed underground canals. Qanats are

BOX 11
Using traditional water conservation and harvesting techniques

Contour bunds are an example cited by Kerr and Sanghi (1992) illustrating a conflict between indigenous and introduced practices. Soil scientists and engineers recommend that bunds be located on the contour so that runoff water is spread evenly. The bunds can reduce runoff, increase infiltration, and divert excess runoff to a central waterway. Most dryland farmers in India have rejected this practice because they want the bunds to conform to field boundaries, which rarely correspond to contours.

Li Shengxiu and Xiao Ling (1992) discussed many indigenous soil- and water-management practices in the drylands of China. The most prominent practices included terracing, frequent shallow cultivation for water conservation, and soil-fertility management. In Gansu Province, "stone fields" are used for growing cereals and fruit trees in an area with an annual precipitation of about 200 mm. This water-conservation practice is centuries old. It involves placing stones on the soil surface to drastically reduce evaporation from the soil surface and to collect dew condensing on the stones and flowing to the soil below during the night. This practice is highly labour-intensive and occurs mainly in areas where subsistence farming is a way of life.

still counted as one of the main ways of procuring water for irrigation and agricultural development as well as drinking-water in the drylands and desert areas of the Islamic Republic of Iran (Farshad and Zinck, 1998) and Afghanistan. However, in most cases, qanats are more than just a way of using groundwater. They represent a unique and integrative system illustrating the use of indigenous knowledge and wisdom in sustainable management of land and water resources. In North Africa and the Sahara, many oases are developed by qanat systems called foggara.

COMBINING MODERN WITH TRADITIONAL TECHNOLOGIES
Traditional rainwater-harvesting agriculture can be a valuable practice in increasing crop productivity in the semi-arid region of the Loess Plateau in

China. However, due to the lack of detailed data on precipitation resources in the region, there have been some difficulties in its development there.

In one study (Hong Wei *et al.*, 2005), based on the precipitation data in the last 40 years and topographical maps at 25 observation stations in and around Dingxi County, Gansu Province, China, raster digital elevation models and average annual precipitation databases in the study areas were established using geographical information systems (GISs). By means of interpolation approaches, statistical models and a comprehensive approach including nine methods (inverse distance weighted, ordinary Kriging, Thiessen polygon, multivariate regression, etc.), the spatial and temporal changes in annual precipitation were calculated and analysed comparatively. The annual average precipitation in Dingxi County calculated by the comprehensive approach is 420 mm, and the water deficit of spring wheat is about 226 mm. Therefore, rainwater-harvesting agriculture is feasible in the study area if appropriate harvesting technologies are applied. The annual average precipitation information system, established by raster precipitation spatial databases using optimized methods, can calculate promptly the total quantities and the spatial changes in precipitation resources on any scale in the study areas. This has an important role in runoff simulation, engineering planning, strategy development, and decision-making as well as water management in rainwater-harvesting agriculture.

SUPPLEMENTARY IRRIGATION IN SEMI-ARID REGIONS

The relationship between grain yield and seasonal evapotranspiration shown in Figures 10 and 11 illustrates why supplemental irrigation is so effective in semi-arid regions. There is usually sufficient precipitation to meet the threshold value required for grain production and to produce some grain (Oweis *et al.* 1999). Therefore, additional water added by irrigation can result in a direct increase in grain yield. The focus of any irrigation system should be on maximizing the evapotranspiration component with added water and minimizing losses such as runoff and deep percolation. This is more difficult under semi-arid conditions than under arid conditions because the rainfall in semi-arid regions is more unpredictable and often

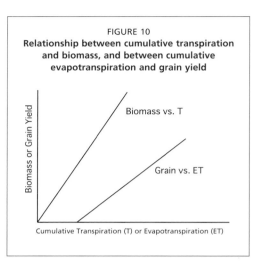

FIGURE 10
Relationship between cumulative transpiration and biomass, and between cumulative evapotranspiration and grain yield

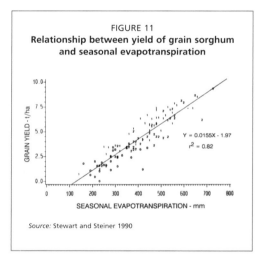

FIGURE 11
Relationship between yield of grain sorghum and seasonal evapotranspiration

$Y = 0.0155X - 1.97$
$r^2 = 0.82$

Source: Stewart and Steiner 1990

ranges from less than half to more than twice the average. Large rainfall events, particularly soon after irrigation, can result in large losses through surface runoff and percolation.

When relatively small amounts of irrigation water are added to grain crops grown under dryland conditions, most of the water will be used for evapotranspiration (Howell, 1990). This is because the soil will be generally dry, so the potential for runoff or percolation of the added irrigation water will be small. However, as more irrigation water is applied, the soil becomes wetter and the potential for losses increases. This is one of the difficulties with efficiently utilizing irrigation water to supplement precipitation. When water resources are limited, it is difficult to determine how much area should be irrigated as a fixed amount of water can irrigate a larger area during a wetter year than during a drier year. Deciding on how much land should be irrigated is critical where water-sensitive crops such as maize are grown. Attempting to irrigate too much land can lead to a water deficiency during a critical growth period such as tasselling (Macartney *et al.* 1971; Rhoads and Bennett, 1991; Oweis, 1997). At the other end of the spectrum, allocating sufficient water to an area so that there will be adequate water for a very high yield even in years of lower than average precipitation can also result in low water-use efficiencies.

Another important factor for grain crops is the harvest index: the ratio of grain weight to the weight of the total above-ground biomass.

Although there are no strategies that can eliminate all these complexities, an understanding of soil–plant–water relationships, precipitation probabilities, and hydrologic characteristics can greatly improve the efficient use of limited water resources for supplemental irrigation. The availability and cost of required infrastructure is perhaps the most important consideration for supplemental irrigation. Even though small amounts of irrigation water can often significantly improve total water-use efficiency, the cost of providing the necessary infrastructure may prevent its use.

CHAPTER 4

Social and economic aspects of Dryland investment

INVESTMENT CONSTRAINTS

Investment to increase production in drylands has been limited, at least in part due to the popular misconception that drylands are empty, barren places (White *et al.*, 2002). Development and research have focused on high-potential areas with the possible expansion of irrigated areas and intensification of irrigated agriculture. However, increasing numbers of people are living in dryland areas as a result of population growth. Hazell (1998) stated that it is becoming increasingly clear that, on poverty and environmental grounds alone, more attention will have to be given by both national governments and international development agencies to less-favoured lands in setting priorities for policy and public investments. At the same time, evidence of productivity gains, poverty reduction, and environmental benefits are required to encourage the necessary funding in dryland regions.

Pender (1999) reviewed the impact of population growth on the degradation or enhancement of soil resources and stated that the evidence is mixed. He cited Tiffen, Mortimore and Gichuki (1994), who found that in the Machakos District in Kenya between the 1930s and the 1990s the population increased fivefold, per capita income increased, erosion was much better controlled, and trees were more prevalent in the landscape. This study supports the optimistic perspectives advanced by Boserup (1965 and 1981) and Ruthenberg (1980), who postulated that households and communities respond positively to pressures induced by population growth, for example by reducing fallow periods, increasing labour and capital inputs per unit of land, developing and adopting labour-intensive technologies, developing more specific property rights, and market development. Pender (1999) concluded that there are many possible household and collective responses to increasing population pressure. These responses are highly site-specific and interact in complex ways. Therefore, he found it difficult to predict the impacts of increasing rural population pressure on the natural-resource base, agricultural production and poverty. However, the impacts of population growth are more likely to be negative where there is no collective response, and positive where population growth induces infrastructure development, collective action, and institutional or organizational development.

There are many examples, particularly in dryland areas, where soil degradation has resulted from population increases. One notable case is the Loess Plateau in China. This region, covering some 53 million ha, is the largest loess area on earth. The plateau is 1 000–1 400 m above sea level and has a loess thickness of 100 m (Wen Dazhong, 1993). The area is characterized by: sloping lands; sparse vegetation; loose soil; and high-intensity, short-duration rainfall. The annual precipitation is 400-600 mm, most of which falls during the summer months. Using historical information, Wen Dazhong (1993) developed an association between population numbers and soil erosion for the Loess Plateau over the last 3 000 years. For many centuries, the erosion rate was relatively low, and this can be attributed primarily to natural processes. Around 1200, the population began to increase and it has grown very rapidly in the past century. The increased food needs of the growing population were met by expanding the

cropland area. Once every piece of flat and fertile land had been used, the inhabitants extended crop cultivation to slopes under natural pasture and forest. Expanding crop production into these areas increased erosion by more than 100 percent, and when soil erosion eventually reduced crop yields on the marginal lands, the people had to destroy more pasture and forests. Forests covered an estimated 40 percent of the Loess Plateau in the period from 221 BC to 581 AD, but only about 6 percent at present. In the worst-affected areas, people collected nearly all the available biomass, including crop residues, leaves and branches of trees, grass roots and manure for household fuel. Removal of this biomass further aggravated soil erosion and reduced the productivity of the land.

Soil conservation in China entered a new era in the 1950s when the Government began to emphasize its importance (Wen Dazhong, 1993). This emphasis is being maintained and erosion has been reduced significantly, even though it remains a major problem. In the Loess Plateau region, about 10 million ha of the eroded area, which account for almost one-quarter of the total erosion in the region, have been controlled since 1950. Terraced fields have been shown to reduce erosion rates by 80 percent, and shrubs and grasses planted on sloping areas can reduce erosion by 70–80 percent. This progress in soil conservation has not only controlled soil erosion, it has also produced economic benefits. For example, crop yields on terraced fields are double those on non-terraced fields, and dam-checked fields have even higher crop yields in some areas. According to an analysis of erosion control practices in small watersheds of this region, the benefit/cost ratio is 2 for erosion control practices, 4 for terraced fields, 1.2 for check-dams, 7 for economic tree plantings, and 10 for shrub plantings (Zhang Jinhui, 1987).

Terraces are effective for conserving soil and water, leading to increased productivity. However, they are costly when large equipment is used and they require large inputs of labour when constructed manually. McLaughlin (1993) reported that terracing of Loess Plateau land in Gansu Province requires 900 labour-days per hectare, not including time for planting crops and for later maintenance. This level of investment is only feasible where land is extremely scarce and the need for food production is paramount. Even so, China terraced more than 2.7 million ha of cropland from about 1950 to 1984, under circumstances that are unlikely to be duplicated elsewhere (Huanghe River Conservancy Commission, 1988). This, coupled with other improved practices, resulted in a 2.8-fold increase in grain production. These extreme measures were required to lessen the widespread hunger in the region, which reached disastrous levels in the late 1950s, when the death rate more than doubled and the birth rate dropped by half (Hellig, 1999). Hunger in the Loess Plateau region was severe with a population of 35.6 million in 1957. By 1981, the population had increased to 72.6 million, doubling in just 24 years (Tian Houmo, 1985).

A significant point illustrated by this example is how rapidly an agro-ecosystem can break down. For centuries, agriculture in the Loess Plateau region was sustainable because of the low population density and the deep loess soils, in spite of significant natural erosion. Then, within a relatively short period, the system began to deteriorate quickly and became clearly unsustainable. However, the example also shows that the downward spiral of degradation can be reversed with institutional intervention, inputs, and infrastructure development.

The International Food Policy Research Institute (IFPRI) reported that land degradation is advancing at an alarming rate in sub-Saharan Africa, particularly as desertification in dryland areas, soil erosion and deforestation in hillside areas, and loss of soil fertility in many cropped areas (IFPRI, 1998). The accelerated soil degradation appears to be related to population increases. While natural forces such as climate change, drought, floods and geological processes contribute to soil degradation, the IFPRI concluded that poverty, rapid population growth and inadequate progress in increasing crop yields were the primary drivers in the land-degradation process. Without substantial investments to improve soil and water management in many areas, conditions will grow worse. The pattern is not homogenous as many issues are site-specific. Investments are required at many levels, including social and institutional investment, applied research, as well as support and incentives to improve soil and water management.

While recognizing that productivity returns had been highest from investments in irrigated and high-potential rainfed lands, Hazell (1998) calls for increased investment in dryland areas. He points out that past decisions had been largely based on the philosophy that adequate poverty reduction and environmental benefits must offset losses in efficiency associated with investing in less-favoured areas. He further states that this view is being challenged by increasing evidence of stagnating productivity growth in many green revolution areas and by emerging evidence that the right kinds of investments can increase productivity to much higher levels than previously thought in some types of less-favoured lands. For example, research in India shows that additional investments in many low-potential rainfed lands can lead to more agricultural growth and a reduction in rural poverty – a "win–win" situation. Hazell concludes that, because of the high levels of investment already made in irrigated and high-potential rainfed areas in India, the marginal returns from some investments (particularly roads, irrigation, education, and agricultural research) are now more attractive in many less-favoured lands. As already discussed, China has been investing on a large scale in the Loess Plateau and other less-favoured lands. Since this effort began in the 1950s, a total of 165 institutes and extension stations for erosion control have been established (Yang Zhenhuai, 1986). In addition, several universities and colleges have established specializations to train soil- and water-conservation professionals. Nevertheless, the development of dryland areas will require investment at many levels before farmers can be successful.

INVESTMENT POTENTIAL

Pingali and Rajaram (1999) stated that there is considerable potential for increasing wheat yields in marginal environments and predicted that future global grain demands could not be met without increasing production in these areas. They emphasized four arguments presented by Byerlee and Morris (1993) to support the allocation of more research resources to marginal environments:

1. Returns to research may now be higher in marginal environments than in more

favourable environments because the incremental productivity of further investment in favourable environments is declining.
2. A large number of people currently depend on marginal environments for their survival, and population pressure is increasing.
3. Because the people who live in marginal environments are often among the poorest groups of the population, increased research investment in the areas is justified on the grounds of equity.
4. Many marginal environments are characterized by a fragile resource base. A special effort is needed to develop appropriate technologies for these areas that will sustain or improve the quality of the resource base in the longer term.

Morris, Belaid and Byerlee (1991) highlighted three related factors that are largely responsible for the relatively slow rate of wheat-yield improvement in marginal environments compared with more favourable areas.
1. The climate in dryland production zones severely constrains the yield potential of cereal crops, so the impact of improved seed and fertilizer technologies is bound to be less dramatic than in the more favourable environments, where these technologies have been highly successful.
2. Investment in agricultural research targeted at rainfed areas has been modest, in part because such research has been perceived as having a lower potential payoff (true only while more responsive alternatives exist).
3. Largely because of the first two, many countries have been slow to implement supportive policies that would promote cereal production in rainfed areas, such as policies to develop market infrastructure.

In contrast, Pingali and Rajaram (1999) point out that wheat research and cultivar development have been fairly successful in these environments despite these factors and contrary to the common perception of the problems of unfavourable environments. Many of the gains in these areas have resulted from the spillover of technologies developed for more favourable environments (Lantican et al., 2003; Dixon et al., 2006). Their findings clearly show that yields have increased

in these areas and that the potential for further increases could be significantly greater if research were aimed specifically at marginal areas.

Perhaps the most compelling reason for increasing investment in dryland areas is the fact that the development of additional irrigation is becoming increasingly difficult and, in many semi-arid regions of the world, irrigated agriculture alone will not be able to satisfy the future demand for food. For example, irrigation development in sub-Saharan Africa is very expensive. Investment costs per hectare in World Bank-funded irrigation projects average about US$18 000, more than 13 times the South Asia average (AQUASTAT, 2008). Moreover, external support for investment in agriculture has declined considerably in the last 20 years.

In sub-Saharan Africa, many countries are classified as "economically water scarce", meaning that they do not currently have adequate infrastructure and capacity (human and financial) to take advantage of the available water resources. Thus, they are not able to cope with the development of irrigation for increased food production and are relying increasingly on the participation of the resource-poor farmers despite their limited access to credit. In other cases, the majority of easily available and inexpensive water resources have been developed.

OPPORTUNITIES AND RISKS OF GROWING FEEDSTOCKS FOR BIOFUELS IN DRYLANDS

Biofuels have been grown and used in drylands for millennia. Twenty-first century interest in biofuels is not in the traditional biofuels (wood fuel and charcoal), but new "generations" of biofuels – principally liquids produced as purportedly environmentally friendly alternatives to petrol and diesel for transport fuels. Ethanol can be produced from sugar cane and cereal crops (currently mainly maize, wheat and sorghum, others are likely to be developed in future), and biodiesel from oil seeds (inter alia oil palm, jatropha, soy, sunflower seeds, coconut and rapeseed) – so called "first generation" fuels. Scientists are working to produce "second generation" fuels, which involve more complex chemical and biological processes using maize stalks, wood waste and by-products of other food and fibre processing.

Use of these recently captured sources of carbon for fuels can bring huge benefit, as they should reduce dependence on fossil fuels. However, as most of the 1st generation biofuels use feedstocks which traditionally enter the human food supply, this raises great concern about the impact which a growing demand for biofuels will have on food supplies and security. Further, devoting land to growing crops and feedstock for 2nd generation biofuels may result in the high risk that energy security could lead to water shortages for agriculture, human populations and food insecurity across the world.

The issues are perhaps most acute in drylands, where food supplies are limited. Traditional food and fibre use of land may lose out in this competition because, on the margin, the potential market for energy is huge and could eventually lead to rising food prices. The latter may not dent the welfare of those who can afford to pay higher prices for both food and fuel, including the population groups that benefit from the development of biofuels. However, low income consumers that do not participate in such gains may be adversely affected in their access to food. Several recent economic studies indicate that increased production of biofuels could lead to price increases not only of crops used for biofuels, but also of other crops – as land is shifted towards greater production of crops for biofuels production. The commercialization of cellulosic-based ethanol (2nd generation fuels from wood and agriculture and forestry waste) could alleviate price pressures while also giving farmers new sources of income, as it would open up new land (like low value grazing lands) to crop production and enable greater productivity from existing cropland (e.g. through use of crop residues for biofuels production) (IEA, 2005).

In the short term, care is required in where, when and how much agricultural land is converted to the growing of crops and feedstock for biofuels to avoid exacerbating food shortages (Worldwatch, 2006), where any transfer of food-growing hectares to biofuel may lead to malnutrition and starvation.

Countries including India have set up policies prohibiting the use of agricultural areas for biofuel production, instead encouraging the use of unused and marginal areas and by-products (of agriculture, food processing and forestry) instead of cereals (Worldwatch, 2006).

With the development and spread of production of 2nd generation fuels, which utilise "waste" streams, crop stems and stover and woody materials by-products, the pressures will be less intense – although this may result in progressive degradation of soil fertility and physical structure (especially due to diversion of sources of organic matter, traditionally returned to the land).

The development of cropping for biofuels (maize and sorghum for ethanol; jatropha for biodiesel; organic "waste" streams for second generation biofuels) in drylands should bring benefits, reducing dependence on imported fuels and improving local access to energy. These potential benefits must be reviewed against potential negative trade-offs, relating to food supply, security and water supplies and regimes. The implications of this use of water in drylands may have deleterious impacts on downstream water users.

In the longer term, the position is more optimistic. Forecasts of world food supply from 2030 to 2050 by FAO, (2006) predict a slowdown in the growth of world agriculture, as world population numbers stabilise. FAO predicts that this slowdown may be mitigated if the use of crop biomass for biofuels were to be further increased and consolidated. Were these to happen, the implications for agriculture and development could be significant for countries with abundant land and climate resources that are suitable for the feedstock. Several dryland countries in Latin America, South-East Asia and sub-Saharan Africa, including some of the most needy and food-insecure ones, could benefit (FAO, 2006).

Successfully planned development of dryland agriculture to produce feedstocks for biofuels offers the opportunity for states to reduce dependence on imported oil for their own fuel needs – and potentially develop an export market for processed biofuels, generating foreign exchange.

PAYMENTS FOR ENVIRONMENTAL SERVICES

Recent years have seen considerable interest in using Payments for Environmental Services (PES) to finance conservation. PES programs seek to capture part of the benefits derived from environmental services (such as clean water) and channel them to natural resource managers who generate these services, thus increasing their incentive to conserve them. Latin America has been particularly receptive to this approach. PES programs are in operation in Costa Rica, Colombia, Ecuador, Mexico and others are under preparation or study in several countries (Pagiola, 2005). The central principles of the PES approach are that those who provide environmental services should be compensated for doing so and that those who receive the services should pay for their provision, also providing additional income sources for poor land users, helping to improve their livelihoods. Several countries are already experimenting with such systems, many with World Bank assistance.

Some hydrologically sensitive watersheds may have very few downstream water users, and so little potential for being included in a PES program. Further, even if poverty rates in target watersheds are high, it does not follow that payments will be received solely, or even principally, by the poor. Even in watersheds with high poverty rates, some land users are likely to be better off, and there can be substantial variability in the level of poverty among the poor.

The potential impacts of PES programs will only be realized by those who participate in the program. Most such programs are too recent for an assessment of participation decisions. But, insights into the factors that are likely to play an important role can be gleaned from the substantial literature that examines the determinants of participation in reforestation, land conservation, and other rural programs (Pagiola, 2005).

CARBON TRADING

The key element of soil rehabilitation in drylands is the restoration of organic matter which has been widely depleted due to tillage, overgrazing and deforestation (Chapter 3), clearly an example of carbon sequestration. The Clean Development Mechanism of the Kyoto Protocol does not include the possibility of payments for carbon sequestration in soils (the Marrakesh Accords established that afforestation and reforestation would be eligible as project based activities) although techniques such as conservation agriculture increase the soil's ability to sequester carbon (Stern, 2006) However, other markets in carbon are being developed, which could enable developing countries to benefit from carbon trading for soil organic matter. By June 2006, the Chicago Climate Exchange (www. chicagoclimateexchange.com) was supporting 350 000 acres of conservation tillage and grass plantings in four states in the USA – acting as a possible model for expansion to benefit projects in drylands of developing countries. Plan Vivo was created by the Edinburgh Centre for Carbon Management in 1996, as a participatory planning and project monitoring system for promoting sustainable livelihoods in rural communities through the creation of verifiable carbon credits. The Plan Vivo System is being applied in Mexico, Mozambique, Uganda and India to generate verifiable carbon credits for sale on the voluntary market and, potentially, for eligibility under the Clean Development Mechanism (CDM within the Kyoto Protocol). All these projects have carbon credits available for purchase. Through the Plan Vivo System, organisations, companies and individuals can not only help offset some GHG emissions but also can help communities in developing countries invest in their own future, while protecting biodiversity, soil and water quality.

ECONOMICS OF WATER HARVESTING

The success of any agricultural development practice ultimately depends on whether or not it is economically, environmentally and socially sustainable. Critchley and Siegert developed a detailed FAO manual for the design and construction of water-harvesting schemes for plant production (FAO, 1991). The technical aspects of water and soil requirements, rainfall-runoff analysis, water-harvesting techniques and crop husbandry were covered in great detail, and there was some discussion of socio-economic factors.

Oweis, Prinz and Hachum (2001) estimated costs of typical water-harvesting practices for

TABLE 8
Costs of water from water harvesting used for crop production

Water harvesting development cost [c]	Life of treatment	Cost per cubic metre of water used for crop production		
		Annual rainfall [a]		
(US$/ha)	(years)	150 mm	300 mm	450 mm
300 (20%) [b]	2	US$0.58 [c]	US$0.29	US$0.19
400 (30%)	3	US$0.36	US$0.18	US$0.12
800 (50%)	4	US$0.34	US$0.17	US$0.11
1 500 (70%)	8	US$0.27	US$0.13	US$0.09
5 000 (90%)	16	US$0.47	US$0.24	US$0.16

[a] 1 mm irrigation supply or rainfall is equivalent to 10 m³/ha.
[b] Numbers in parentheses are the percentages of rainfall harvested by the land treatment applied to induce runoff.
[c] Development cost per hectare divided by number of cubic metres per hectare delivered annually (cubic metres per hectare of annual rainfall multiplied by proportion harvested), multiplied by the appropriate annuity at a 10-percent discount rate (2 years US$0.5764 per US$1 invested; 3 years US$0.4023; 4 years US$0.3156; 8 years US$0.1875; 16 years US$0.1278).

inducing runoff in countries of the Near East. While these values are not necessarily applicable to other areas, they do provide a basis for making some estimates of benefit/cost ratios for water harvesting compared with irrigation development. On the basis of these estimates, the costs for harvesting 1 m³ of water were estimated for areas with different annual rainfall (Table 8). They were estimated as annuities, assuming a 10 percent discount rate over the lifetimes of different treatments of the runoff catchment areas. The costs ranged from US$0.09 to US$0.58

per cubic metre of water, depending on rainfall and the cost-effectiveness of treatments.

For comparison, Table 9 lists development costs for delivering 1 m³ of water by different irrigation systems. Estimated irrigation development costs ranged from US$0.03 to US$0.33 per cubic metre of water. They depend on the cost of developing the infrastructure and on how much the system is used. Where irrigation sources are not limited and climatic conditions allow the growing of two or more crops per year, so that 1 000 mm of

TABLE 9
Development costs of water from irrigation systems used for crop production

Irrigation development cost [b]	Life of system	Cost per cubic metre of water used for crop production		
		Irrigation supply [a]		
(US$/ha)	(years)	500 mm/year	750 mm/year	1 000 mm/year
2 500	25+	US$0.06 [b]	US$0.04	US$0.03
5 000	25+	US$0.11	US$0.07	US$0.06
7 500	25+	US$0.17	US$0.11	US$0.08
10 000	25+	US$0.22	US$0.15	US$0.11
15 000	25+	US$0.33	US$0.22	US$0.17

[a] 1 mm irrigation supply or rainfall is equivalent to 10 m³/ha.
[b] Development cost per hectare divided by number of cubic metres per hectare delivered annually, multiplied by the 25-year annuity at a 10-percent discount rate (US$0.1102 per US$1 invested). Operation and maintenance costs not included.

TABLE 10
Development and total costs of water used from shallow and moderately deep small tube wells

Unit cost	Lifetime	Lift	Discharge		Investment [b]	Operation [c]	Maintenance [d]	Total cost
(US$)	(years)	(m)	(litres/s)	(m³/year) [a]	(US$/m³)			
2 500	4	5	2	7 200	0.11	0.03	0.05	0.19
2 500	4	2	5	18 000	0.04	0.01	0.02	0.08
5 000	8	25	2	7 200	0.13	0.10	0.07	0.30
5 000	8	10	5	18 000	0.05	0.04	0.03	0.12

[a] Assuming 1 000 operation hours per year.
[b] Investment cost per tubewell divided by volume (cubic metres per year, multiplied by the appropriate annuity at a 10-percent discount rate (4 years US$0.3156 per US$1 invested; 8 years US$0.1875).
[c] Shallow tubewell 3 kW, US$0.07/kWh; deeper tubewell 15 kW, US$0.05/kWh.
[d] Shallow tubewell 15 percent of unit cost/year, deeper tubewell 10 percent/year.

irrigation water can be used, the cost per cubic metre is relatively low even when the cost of development is high. The development cost per cubic metre increases rapidly when the system provides only 500 mm water or less per year. The development cost per cubic metre water used was estimated as an annuity, assuming a life expectancy of 25 years for the systems and a 10 percent discount rate. These estimates do not include operation and maintenance costs – and are clearly academic where farmers do not have access to credit. Table 10 lists estimated development and total costs per cubic metre of water for shallow and moderately deep small tubewells, on the basis of similar assumptions.

Although irrigation development is usually not an option in areas where water harvesting is practised, it is of interest to compare their development costs. The estimates in Tables 8 and 9 suggest that water harvesting is less cost-effective than irrigation even when the irrigation development cost is US$10 000–15 000/ha. This is particularly true in the lower-rainfall regions. Oweis, Prinz and Hachum (2001) consider that most enhanced water-harvesting catchments are short-lived (Table 6). Thus, although the initial cost is high, the most cost-effective water-harvesting practice in the long term may be to use an impermeable cover such as plastic or asphalt so that a high proportion of the precipitation can be harvested (details in Tables 5 and 6). The estimated development costs are only for harvesting the water and having it run onto adjacent cropped land as a supplement to the rainfall. Where

the water is intended for use as supplementary irrigation, a tank or reservoir must be available or installed to store the water until it is needed. In addition, a pump or other means of delivering the water to the crop might be required. This would entail substantial additional costs, and there may also be considerable loss from evaporation or seepage during storage.

Even in drylands, there may be opportunities to develop local, small-scale groundwater resources. Therefore, a comparison of water-harvesting costs with those of small tubewells may be more relevant than with those of irrigation systems. A comparison of Tables 8 and 10 shows that the total costs of water generated through runoff agriculture (without additional water-storage facilities) are of the same order as water costs from shallow or moderately deep tubewells in areas with rainfall of about 450 mm. In lower-rainfall areas, tubewells would be more economic where shallow groundwater of adequate quality were available. Individual catchment areas in many types of runoff farming are smaller than 1 ha, with some much smaller. Consequently the comparison with tubewell-irrigated agriculture, although more direct than with large-scale irrigation, is still across different farming systems.

Perhaps the greatest problem in assessing the benefit/cost ratio of water-harvesting practices is the lack of assurance that water will be available. This is true particularly where perennial crops or trees depend on harvested water. One approach is to include both perennial crops and annual grain crops in the design. When rainfall is adequate, supplemental irrigation water will be available for both. In dry years, water will only be available for the perennial crops, but the design would be conservative enough to almost guarantee sufficient water to maintain the perennial crops even during prolonged droughts.

Supplemental water can be very beneficial when it is available at critical periods (providing application is not immediately followed by heavy rain). Oweis, Prinz and Hachum (2001) showed that 1 m³ of water applied as supplemental irrigation could produce 2–3 kg of wheat. This compared favourably with the productivity of water from full irrigation, which was in the order of 1 kg/m³. Even using the high value of 3 kg of wheat per cubic metre, the costs of harvesting water for cereal production often cannot be justified where only price is considered. The perspective may be more favourable with higher-value crops. There are also other benefits, such as maintaining a local supply of food, making use of available family labour in some cases, enhancing the environment, and other social benefits that might make water harvesting more feasible than on the basis of production economics alone.

Economic considerations suggest that water harvesting is most attractive where the harvested water can be used directly by crops on an adjacent area; next where water can be diverted to nearby crops or trees; and least where the harvested water must be stored and used later as irrigation. Although the potential for water harvesting in dryland regions is considerable, there are many problems in addition to constructing the systems that have constrained wide-scale development of water-harvesting systems. Records on water-harvesting areas are often not definitive, with insufficient data for good designs. In some years, there is not enough harvested water to be successful. In other years, waterlogging may be a problem. Erosion on lands receiving harvested water can also be a difficulty, and the maintenance of water-harvesting systems can be labour-intensive and costly. Nevertheless, water-harvesting systems in dryland regions must be given more focus. They may very well be more economically feasible for growing tree crops or other high-value crops than for grain crops. Analysis and design should

be based on rainfall-probability distributions rather than average values. Probabilities provide a more realistic evaluation than average values because the rainfall amounts in dryland regions are very erratic.

ECONOMICS OF WATER-CONSERVATION PRACTICES

A realistic goal for producers in dryland regions is to increase growing-season evapotranspiration of grain crops by 25 mm. The effect of this on grain yield can be estimated on the basis of grain-yield and water-supply information. Musick and Porter (1990), Rhoads and Bennett (1991), and Krieg and Lascano (1991) reviewed the water-use efficiencies of wheat, maize and sorghum, respectively, which varied considerably depending on yield levels and climate conditions in the many studies conducted worldwide. However, as a general guide, 1.7 kg of maize grain, 1.5 kg of sorghum or 1.3 kg of wheat can be produced in dryland regions per additional cubic metre of water used by evapotranspiration. These values can be refined where sufficient local data are available. Using these values, some preliminary benefit/cost estimates can be made regarding the amount of investment that can be made based only on production. However, there may be social and environmental benefits that will justify investment costs far beyond those strictly for increased grain production.

Based on the water-use efficiency values above, the average yield of wheat could be increased by 0.32 tonnes/ha by an extra 25 mm seasonal evapotranspiration. FAO (1996a) reported that the 1988–1990 average yield of wheat in developing countries in semi-arid regions was 1.1 tonnes/ha, so this would represent a 30 percent yield increase. Maize yield could be increased by 0.42 tonnes/ha, and sorghum by 0.38 tonnes/ha. The 1988–1990 average yields of maize and sorghum in semi-arid regions of developing countries were 1.13 and 0.65 tonnes/ha, respectively (FAO, 1996a). Therefore, increasing plant water use by only 25 mm could potentially raise the average yields of maize and sorghum by 38 and 58 percent, respectively. These large gains from such a small amount of additional water use are feasible because the threshold amounts of water required for grain production are already met (Figure 5), and the additional water increases grain production directly, providing the water is available at the critical period of the growing season and that sufficient plant nutrients are available to take advantage of the additional water use. In some cases, this will imply the addition of organic matter or mineral fertilizers.

The most effective system for conserving water is no-tillage farming (Rockwood and Lal, 1974; Scoones et al., 1996; Tebrügge, 2000; FAO, 2001b; FAO, 2001d). Its effectiveness has been proven in many areas (see Annex 2), but as already discussed, its adoption has been limited because of the cost of new or modified tools, equipment or other inputs, and the high level of management required (Benites & Friedrich, 1998). No-tillage is also disregarded by many producers in developing countries because of conflicting demands on crop residues for animal feed or for household fuel. In situations where no-tillage or another form of conservation agriculture is not feasible, terracing or land levelling may be required to prevent runoff in order to increase the amount of plant-available water. The specific practices will depend on social and economic conditions, soil and terrain variables, and climate.

The benefits from conservation agriculture accrue slowly. Because of this, producers may become disappointed and abandon the systems after only a few years. Several years are required to enhance soil porosity and organic matter content to the point that significant yield increases are apparent (FAO, 2001d; WOCAT, 2007). Although it takes several years to increase the soil organic matter content significantly by conservation agriculture, the increases can be lost in a very short time by just one intensive tillage operation (Fowler and Rockström, 2001; Mrabet, 2002). Therefore, once a producer adopts conservation agriculture, it is important that every effort be made to continue.

Conservation agriculture practices can make better use of the limited amounts of precipitation in dryland areas and increase yields significantly (Mortimore, 1998; FAO, 2001d; Lal, 2002b). This can be an important step towards increasing cereal production and improving the well-being

of people living in dryland regions. However, improving the yields in dryland areas by 30–50 percent will have only a relatively small effect on global cereal production. FAO (1996a) reported that only about 10 percent of the wheat, 8 percent of the maize, and 35 percent of the sorghum produced in developing countries were grown in semi-arid regions. Therefore, even though it is highly important to increase investment in dryland regions, agriculture in the more favourable climate regions and irrigated areas must continue to become more efficient if food and fibre supply are to keep pace with population growth.

CURRENT SCENARIO IN DRYLAND REGIONS

In the present circumstances, dryland farming is a risky enterprise. Drought is the principal hazard facing dryland farmers but insects, hail, intense torrential rains and high winds can also damage or destroy crops. Little can be done to prevent most sudden disasters, but there are soil- and crop-management practices that can reduce the impact of all but the most protracted droughts. While low soil-water content commonly restricts crop yields in dryland areas, there are several other soil problems (surface-soil hardening, compaction, water and wind erosion, low soil fertility, shallow soils, restricted soil drainage and salinization) that can also affect dryland production (Singh, 1995).

Improved management of land and water resources can counter desertification (whether due to climate change and / or overgrazing & deforestation) and increase the productivity of low-rainfall areas (Steiner et al. 1988). Lal (1987) presented a review of available low-input technologies that can improve the productivity of dryland regions and protect their soil resources from erosion. In order to ensure that the true causes of the problem of drought are identified and that potential solutions are feasible and acceptable to the farmers, a participatory approach must be adopted. Farmers testing possible solutions on their own farms are also encouraged to be more innovative. This is particularly important because farmer innovation is the key to the sustainability of the agricultural development process, especially in the situation all too common in developing countries, with inadequate advisory services.

Achieving long-term sustained growth in the productive capacity of low-rainfall areas requires sound decisions based on accurate assessments of resource problems and potentials, combined with careful analyses of alternative policies, programmes and projects. A study by FAO (1986) outlined specific practices and policies needed to improve African agricultural productivity, focusing on the provision of incentives, inputs, institutions and infrastructure.

General development goals for improving and sustaining the productivity of dryland areas include:
- improvement in the livelihoods of people living in dryland areas;
- a shift from conventional agriculture to a more conservation-effective agriculture (i.e. adoption of agro-ecosystem approaches and conservation agriculture);
- a greater contribution of dryland regions to the growth and development of national economies;
- a sustained productive life of drylands by arresting the processes of land degradation;
- rehabilitation of seriously degraded land;
- adoption and spread of dryland-management systems that are economically and socially viable and environmentally sustainable;
- improved decision-making abilities of local, national and regional (e.g. river basin level) planners.

A more conservation-effective agriculture should be promoted in response to the decline in land productivity under conventional agricultural systems-and to mitigate the effects of climate change. Conventional practices of particular concern include: continuous cultivation using mould-board ploughs or other intensive tillage tools; removal or burning of crop residues; inadequate rotations or monoculture; and overgrazing and deforestation that do not maintain vegetative soil cover or allow appropriate restitution of soil organic matter and plant nutrients.

The strategy for conservation agriculture has four components:
- using no-tillage or minimum-tillage systems;
- maintaining soil cover at all times;
- using suitable crop rotations;
- integrating livestock with cropping systems.

This strategy minimizes soil disturbance, enhances vegetative cover and contributes to the sustained use of agricultural soils. An effective participatory approach to research and extension is needed for the successful adoption of conservation agriculture. The "win-win" impacts of conservation agriculture include:
- labour savings and reduced peak labour demand;
- improved soil organic matter content and biological activity;
- improved soil structure and moisture availability;
- reduced erosion and runoff;
- improved crop yields (totals and reliability);
- crop diversification;
- increased income opportunities.

CHAPTER 5

Wider issues of water in Drylands

Water harvesting (WH) and soil water conservation (SWC) in croplands and on pasture hold the potential to contribute to the vital wider development of drylands pasture by increasing the yields and their reliability. If well planned, successful WH and SWC initatives can create the conditions required to enable local land users (smallholders, agropastoralists and pastoralists) to escape from the vicious cycles that lead to land degradation and rural poverty, by contributing to poverty reduction and economic growth. However, there are potential negative trade-offs associated with increased agricultural water use, which should be anticipated and plans made to limit potentially deleterious impacts.

Watersheds (river basins) have long been acknowledged as the appropriate and logical unit of analysis and planning for improving water resources. Through watershed management, it has been proven that soils and water (surface and ground) resources can be better managed and sustained using SWC (Kerr, 2002; WRI, 2005; Brooks and Tayaa, 2002; WOCAT, 2007). The watershed approach encourages the promotion of co-operation between upstream and downstream stakeholders – in an effort to minimize conflicts over land and water. Plans must make sense both economically and environmentally – to contribute to poverty reduction and improve the functioning of the

watersheds – particularly to restore the recent widespread reduction in groundwater levels in semi-arid areas (Seckler, 1998). The watershed approach also provides clarity in determining the economic importance of water-related ecosystem service (e.g. increasing water yield, improving water quality, reducing sediment delivery to a reservoir). The approach can be used at a range of spatial scales, from micro-catchments upwards.

Kerr (2002) reviewed the outcomes of a large number of watershed development projects in India, all designed to realize hopes for agricultural development in rainfed, semi-arid areas. These areas were bypassed by the Green Revolution and had experienced little or no growth in agricultural production for several decades. The case studies in Andhra Pradesh and Maharashtra, India offer important insights for other parts of the world. By systematically evaluating the opportunities and challenges of watershed development, Kerr concluded that while most of the projects they surveyed have had relatively little impact, those that take a more participatory approach and are managed by NGOs have performed better in conserving natural resources and raising agricultural productivity. The author cautioned that success often comes at the expense of the poorest people in watershed areas; improving the management of a watershed usually requires restricting access to the natural resource base on which they depend. Many watershed development projects do not work because those whose interests are harmed refuse to go along with the effort. The author argued that for watershed development to succeed on a large scale, projects must find a way for all affected parties to share in the net gains generated.

A detailed analysis of the benefits of one of the projects (part on an Indo-German Watershed Development Program) in drought-plagued Maharashtra, India around Darewadi Village (WRI, 2005) demonstrates the dramatic success possible with careful planning. In 1996, the main village and its twelve hamlets were on the verge of desertification. Before the watershed was regenerated Darewadi's 921 residents depended on water deliveries from a tanker truck for four months per year. In 2004 the village was tanker free, despite receiving only 350mm of rain in 2003 – 100mm less than its annual average. The program at Darewadi involved five years of regeneration activities, including tree and grass planting, a grazing ban, sustainable crop cultivation (decreasing the need to purchase inorganic fertilizers – which are energy expensive to produce), soil and water conservation measures, construction of simple water harvesting and irrigation systems (hillside contour trenches and rainwater harvesting dams). The grazing restrictions were lifted after five years, livestock number rebounded depending on more plentiful fodder and yields (milk and crop) increased. Signs of increased household wealth and well-being appeared.

In reviewing the high level of attention being given to water harvesting and groundwater recharge in Rajasthan between 1974 and 2002, when the state government alone invested 8 534 930 000 rupees (approximately US$190 million) in watershed treatment, Rathore (2005) was unable to locate any systematic scientific evaluation regarding the effectiveness of recharge techniques. This should not be interpreted as indicating that water-harvesting efforts themselves have had little impact. Rather, it simply indicates that available technical evaluations are inadequate to reach any conclusion, meaning that potentially valuable lessons could not be learned.

In a wider review of watershed rehabilitation across India, Saxena (2001) noted evaluation reports showed that watershed rehabilitation will fail to meet productivity, equity and sustainability objectives unless project beneficiaries are fully engaged and careful attention is paid to issues of social organisation. Success depends on consensus among a large number of users and collective capability is required for management of the commons, also of new water harvesting structures created during the project. He concluded that the costs and benefits of watershed interventions will be location-specific and, concurring with Kerr (2002), unevenly distributed among the people affected especially where poorer groups are unable to have their requirements met.

The record of government agencies in stimulating people's participation has been poor and their overall success rate low (Kerr *et al.*, 2000). Field staff were found to have no incentive to make the effort to pursue participatory approaches. Saxena (2001) concluded that lands in the upper catchment should be rehabilitated first for at least three reasons. First, so that the landless and the poor who depend on the upper slopes can benefit; second, so that groundwater recharges as early as possible; and third, by the time the lower catchment is treated any debris and erosion running down from the upper catchment has been minimised. High priority should also be given to rejuvenation of village ponds and tanks, and recharge of groundwater. Despite problems there are many success stories, especially in States such as Madhya Pradesh and Andhra Pradesh. Successful and sustainable projects such as Ralegaon Siddhi, the revival of johad in Alwar, Sadguru's activities in Gujarat, and watershed development in Jhabua and Sagar districts of Madhya Pradesh have characteristics which include: the emphasis on social issues, people's mobilisation, clear direction to Government machinery to accept principles of participatory management, explicit project monitoring and a strong sense of ownership by the local community.

The insights from these analyses of watershed management projects (Kerr 2000 & 2002; Saxena, 2001; WRI, 2005; Rathore, 2005) should be used as lessons to guide developments to improve water use and governance in other drylands. These models can be locally adapted to help restore groundwater, increasing crop and pasture yields, also reducing the energy required to pump water for household crop-use- contributing to poverty reduction and economic growth, particularly vital in Africa.

The two major environmental implications of the tremendous increase in the use of inorganic fertilizers are the energy costs of production and distribution, also the impact of the fertilizers on groundwater (Oberthür and Ott, 1999). SWC projects restore the organic matter content of dryland soils and raise fertility levels, reducing the need to use inorganic fertilizers – and consequently reducing energy use.

Success in water harvesting and soil water conservation, diverting a greater proportion of the precipitation for crop and pasture growth (green water) can risk directly reducing the availability of water downstream (blue water) – for urban areas, irrigation and reservoirs. One of the most conspicuous results of overuse of water harvesting and irrigation is that some large rivers now dry up before reaching the sea. Increased water use in one area may entail reduced availability in another downstream It is vital to get people involved in water management for agriculture at local level by real participation and transparent decision making. What is proposed is a new water contract. The Green Revolution was staged by scientists, the Blue Revolution should be staged by making water use and management everyone's business: its goal would be to maximize the production of food and the creation of jobs per water unit consumed. Enabling individuals and communities to understand their options for change, to choose from these options, to assume the responsibilities that these choices imply, and then to realize their choices could radically alter the way the world uses its limited water resources. The ultimate aim of water management is to optimize water use throughout a river basin in such a way that all users have access to the water they need (FAO, 2002).

Achieving sustainable agro-ecosystems is a major challenge of the current century. With increasing populations and improved living standards, the demand for food and fibre will drive the development of sustainable and more productive agro-ecosystems, particularly in less favourable regions, including drylands. This challenge will become substantially greater if the recent trend of using cereals for producing ethanol for fuel continues to expand and encourages cereal production into even less favourable regions. There is an imbalance between natural resources, population and basic human needs in many places, particularly in semi-arid regions. Agro-ecosystems in these areas can be developed and sustained, but careful management will be required and productivity will be low and highly variable even when the best technologies are used.

The prevention of soil degradation is the first and most important issue that must be addressed in dryland agriculture. Soil organic matter is correlated significantly with soil water-holding capacity, fertility and productivity. Therefore, maintaining and increasing SOM is of critical importance. A permanent cover on the soil reduces or eliminates runoff and erosion, reduces soil surface temperatures and thus slows down the decomposition of organic matter. Tillage drastically increases the rate of decomposition. Therefore, it is imperative that tillage be reduced or avoided and that cover be maintained on the soil surface in dryland cropping systems if soil degradation is to be reduced or reversed. This is a tremendous challenge in semi-arid regions because insufficient precipitation seriously limits organic matter production. Moreover, the generally warm conditions accelerate the decomposition of in situ SOM during periods of favourable soil-water conditions. The challenge is even greater in many developing countries where crop residues are removed for livestock feed or household fuel – and in future may also be lost to the agro-ecosystem through use to produce biofuels.

Improved water management is the other key factor that must be addressed in dryland regions. Where water is severely limited, other technological advances such as improved varieties, fertilizers and pesticides, are generally not effective. Small increases in seasonal water use by plants can increase yields significantly. It is estimated that increasing seasonal plant water use by 25 mm, a realistic amount that can be gained through improved water management, could increase the average yields of wheat, maize and sorghum in developing countries by 30, 38 and 58 percent, respectively.

Water harvesting is also a promising practice for some situations. It is most attractive where water can be used directly by crops on an adjacent area. However, water harvesting also holds promise in some cases where harvested water can be stored in cisterns, ponds and other places for later use. This is because small amounts of water applied at critical growth stages can be highly effective.

A considerable body of research knowledge and producer experience exists. This information is sufficient in many cases for developing sustainable agro-ecosystems. Therefore, the greatest challenge is the implementation and execution of sound management plans. Such management plans must incorporate due consideration of the:
- impacts which climate change will inevitably have on dryland agro-ecosystems;
- methods of mitigating unintended impacts (downstream) of water harvesting and agricultural intensification;
- potential impacts of the developing biofuels industry, both beneficial and detrimental, as they raise new pressures on delicate dryland agro-ecosystems.

The techniques outlined in this text form a range of options for improved technologies and management practices, some of which will be appropriate in some situations and not in others. Some options may be 'stand-alone', others complementary so could be beneficially used in a sequence to provide a locally appropriate solution. It is not possible to offer any ranking to suggest one option is better than another; such a judgement should only be made with the benefit of local knowledge, in participation with local land users (small-holders, agropastoralists and/or pastoralists). Lessons learned, particularly from South Asia, demonstrate that local participation and understanding are vital for success where watershed approaches are adopted.

Sustainable systems must focus on long-term goals, but the reality is that short-term benefits and solutions almost always take precedence over long-term issues. Historically, agro-ecosystems have been developed for short-term benefits without a thorough analysis of the long-term consequences. Lessons learned, particularly in areas of South Asia, indicate that integrated watershed management approaches can contribute to poverty reduction and economic development. These provide models for developments elsewhere, notably in Africa; offering ways to reduce the fertilizer, labour and energy demands of agriculture, while sustaining or raising crop and pasture production. Producers, scientists, policy-makers and governments must work closely together to produce adequate amounts of food and fibre and meet the challenge of sustaining the natural-resource base.

Al Ghariani, S.A. 1995. Supplemental irrigation and water harvesting systems in Libya. *In* E.R. Perrier & A.B. Salkini, eds. *Supplemental irrigation in the Near East and North Africa*, pp. 425–447. Dordrecht, The Netherlands, Kluwer Academic Publishers.

AQUASTAT. 2008. Database on investment costs of irrigation. Food and Agriculture Organization of the United Nations, Rome. (available at **http://www.fao.org/nr/water/aquastat/investment/index.stm**).

Armitage, F.B. 1987. Irrigated forestry in arid and semi-arid lands: a synthesis. International Development Research Center, Ottawa.

Arrouays, D. & Pélissier, P. 1994. Changes in carbon storage in temperate humic soils after forest clearing and continuous corn cropping in France. *Plant Soil*, 160: 215–223.

ASA. 1976. *Multiple cropping*. ASA Special Publication No. 27. Participating agencies: Crop Science Society of America. Soil Science Society of America. Madison, USA, American Society of Agronomy.

Australian Centre for International Agricultural Research. 2002. Improving water-use efficiency in dryland cropping. Canberra. (also available at **www.aciar.gov.au**).

Baker, C.J, Saxton K.E, Ritchie, W.R., Chamen, W.C.T., Reicosky, D.C., Ribeiro, M.F.S., Justice, S.E. & Hobbs, P.R. 2007. *No-tillage seeding in conservation agriculture*. Food and Agriculture Organization, Rome and CAB International , Wallingford, UK

Bamatraf, A.M. 1991. Supplemental irrigation in Yemen Arab Republic. *In* E.R. Perrier & A.B. Salkini, eds. *Supplemental irrigation in the Near East and North Africa*, pp. 561–598. Dordrecht, The Netherlands, Kluwer Academic Publishers.

Bandaru, V., Stewart, B.A., Baumhardt, R.L., Ambati, S., Robinson, C.A. & Schlegel, A. 2006. Growing dryland grain sorghum in clumps to reduce vegetative growth and increase yield. *Agronomy Journal* 98:1109-1120.

Batjes, H.N. & Sombroek, W.G. 1997. Possibilities for C sequestration in tropical and subtropical soils. *Global Change Biology* 3:163-173.

Batchelor, C., Singh A., Rao, R.M., and Butterworth, J. 2002. Mitigating the potential unintended impacts of water harvesting. IWRA International Regional Symposium on Water for Human Survival, 26-29 November, 2002, Hotel Taj Palace, New Delhi, India.

Ben–Asher, J. 1988. A review of water harvesting in Israel. Draft working paper for World Bank's Sub–Saharan Water Harvesting Study. Washington, DC, World Bank.

Benites, J. & Friedrich, T. 1998. Overcoming constraints in the adoption of conservation tillage practices. *In*: *Conservation tillage for sustainable agriculture*, Part 2, pp. 219–221 Proc. International Workshop, Harare, 22–27 June 1998. Eschborn, Germany, GTZ Dep. 45 Rural Development.

Borlaug, N. 1996a. Feeding the world: the challenges ahead. *In* A. Dil, ed. *Norman Borlaug on world hunger*, pp. 485–489. San Diego, USA, Bookservice International, and Lahore, Pakistan, Ferozsons (Pvt) Ltd.

Borlaug, N. 1996b. The acid soils: one of agriculture's last frontiers. pp. 467–484 *In* A. Dil, ed. *Norman Borlaug on world hunger*, pp. 467–484. San Diego, USA, Bookservice International, and Lahore, Pakistan, Ferozsons (Pvt) Ltd.

Boserup, E. 1965. *The conditions of agricultural growth*. New York, USA, Aldine Publishing Co.

Boserup, E. 1981. *Population and technological change: a study of long–term change*. Chicago, USA, University of Chicago Press.

Bowden, L. 1979. Development of present dryland farming systems. *In* A.E. Hall, G.H. Cannell & H.W. Lawton, eds. *Agriculture in semi–arid environments*. pp. 45–72. Berlin, Springer Verlag.

Boyce, K.G., Tow, P.G. & Koocheki, A. 1991. Comparisons of agriculture in countries with Mediterranean–type climates. *In* P.W. Unger, T.V. Sneed, W.R. Jordan & R. Jensen, eds. *Challenges in dryland agriculture: a global perspective*, pp. 250–260 Proc. International Conference on Dryland Farming, Amarillo/Bushland, USA, 15–19 August 1988. USA, Texas Agricultural Experiment Station.

Brooks, K.N., Ffolliott, P.F., Gregersen, H.M. & DeBano, L.F. 1997. *Hydrology and the management of watersheds*. Ames, USA, Iowa State University Press.

Brooks, K.N. & Tayaa, M. 2002. Planning and managing soil and water resources in drylands: role of watershed management. Arid Lands Newsletter No. 52, International Arid Lands Consortium.

Brown. L.R., 2008. *Plan B 3.0 Mobilizing to save civilization.* Earth Policy Institute, W.W. Norton & Company, New York.

Burger, W.C. 1981. Why are there so many kinds of flowering plants? *BioScience* 31:572, 577-581.

Burnett, E., Stewart, B.A. & Black, A.L. 1985. Regional effects of soil erosion on crop productivity–Great Plains. *In* R.F. Follett & B.A. Stewart, eds. *Soil erosion and productivity*, pp. 285–304. Madison, USA, American Society of Agronomy, Crop Science Society of America and Soil Science Society of America.

Byerlee, D. & Morris, M. 1993. Research for marginal environments: are we underinvested? *Food Pol.*, 18: 381–393.

Campbell, H.W. 1907. Campbell's 1907 soil culture manual. Lincoln, USA, H.W. Campbell.

CBD. 2007. Convention on Biological Diversity. Montreal, Canada. (available at **http://www.cbd.int/**)

Centre for Science and Environment. 2001. *Making water everybody's business: practice and policy of water harvesting.* (available at **http://www.oneworld.org**).

Chander, K., Goyal G., Nandal, D.P., & Kapoor, K.K. 1997. Soil organic matter, microbial biomass and enzyme activities in a tropical agroforestry system. *Biology and Fertility of Soils* 27:168-172.

China Ministry of Science and Technology. 2001. China reclaiming forests and grassland from farming land. *Chin. Sci. Tech. News.*, 269.

Cooper, J.M.,Gregory, P.J., Tully, D. & Harris, H.C. 1987. Improving water use efficiency of annual crops in the rainfed farming systems of west Asia and North Africa. *Experimental Agriculture* 23:113-158.

Cornish, P.S. & Pratley, J.E. 1991. Tillage practices in sustainable farming systems. *In* V. Squires & P. Tow, eds. *Dryland farming: a systems approach*, pp. 76–101. Sydney, Australia, Sydney University Press.

Cosgrove, W.J. & Rijsberman, F.R. 2000. *World water vision: making water everybody's business.* London, Earthscan Publications Ltd.

Dhruva, N. & Babu, R. 1985. *Soil and water conservation in semi-arid regions of India.* Dehradun, India Central Soil and Water Conservation Research and Training Institute.

Dixon, J., Nalley, L., Kosina, P., La Rovere, R., Hellin, J. & Aquino, P. 2006. Adoption and economic impact of improved wheat varieties in the developing world. *Journal Agricultural Science* 144:489-502.

Dregne, H.E. 2002. Land degradation in the drylands, *Arid Land Res. Manag.* 16:99-132.

Dregne, H. 1982. *Dryland soil resources.* Washington, DC, Agency for International Development, Department of State.

Dregne, H.E. & Willis, W.O. (eds.). 1983. *Dryland Agriculture.* Agronomy No. 23, American Society of Agronomy, Crop Science Society of America, Soil Science Society of America, Madison, WI.

Dregne, H., Kassas, M. & Rozanov, B. 1991. *A new assessment of the world status of desertification.* Desertification Control Bulletin. UNEP.

El Titi, A. 1999. *Lautenbacher Hof Abschlussbericht 1978–1994.* Agrarforschung in Baden–Württemberg, Band 30. Stuttgart, Germany, Ministerium Ländlicher Raum.

El-Swaify, E.S., Pathak, P., Rego, T.J. & Singh, S. 1985. Soil management for optimized productivity under rainfed conditions in the semi-arid tropics. *Adv. Soil Sci.*, 1: 1–64.

Erskine, W. & Malhotra, R.S. 1997. Progress in breeding, selecting and delivering production packages for winter sowing chickpea and lentil. *In: Problems and prospects of winter sowing of grain legumes in Europe*, pp. 43–50. AEP Workshop, Dijon, December 1996. France, AEP.

Falkenmark, M. 1995. Integrated land and water management: challenges and new opportunities. *Ambio*, 24:1, 68.

FAO. 1986. *African agriculture: the next 25 years.* Rome.

FAO. 1987. *Soil and water conservation in semi-arid areas.* (available at **http://www.fao.org**).

FAO. 1991. A *manual for the design and construction of water harvesting schemes for plant production.* Rome.

FAO. 1995. *Irrigation in Africa in figures.* Water Report No. 7. Rome.

FAO. 1996a. *Food production: the critical role of water.* Background paper for the World Food Summit. Rome.

FAO. 1996b. Prospects to 2010: agricultural resources and yields in developing countries. *In: Volume 1, Technical background documents*, 1–5, pp. 26–36. World Food Summit. Rome.

FAO. 1998a. *Predictions of cattle industry, cultivation levels, and farming systems in Kenya* by W. Wint & D. Rogers. AGA Consultants' Report. Animal Production and Health Division. Rome.

FAO. 1998b. *The role of green water in food production in Sub-Saharan Africa*, by H.H.G. Savenije (available at **http://www.fao.org**).

FAO. 2000a. *Land resource potential and constraints at regional and country levels.* World Soil Resources Report No. 90. Rome.

FAO. 2000b. *Food security and sustainable development in the Middle East Region,* by P. Koohafkan. First session, FAO Agriculture, Land and Water Use Commission for the Near East (ALAWUC), 25–27 March 2000, Beirut. (also available at **http://www.fao.org**).

FAO. 2000c. Carbon sequestration options under the clean development mechanism to address land degradation. *World Soil Resources Report* No. 92. Rome.

FAO. 2001a. *Global agro-ecological zones.* Rome (also available at **http://www.fao.org**).

FAO. 2001b. *The economics of conservation tillage.* Rome.

FAO. 2001c. *Farming systems and poverty. Improving farmers' livelihoods in a changing world*, by J. Dixon, A. Gulliver & D. Gibbon. Rome, and World Bank (available at **http://www.fao.org**).

FAO. 2001d. The importance of soil organic matter: key to drought resistant soil and sustained food production. *FAO Soils Bulletin 80*. Food and Agriculture Organization of the United Nations, Rome.

FAO. 2002. *Land and Agriculture: from UNCED, Rio de Janeiro 1992 to WSSD*, Johannesburg 2002, J. Pretty & P. Koohafkan. Rome.

FAO. 2003a. Compendium of Agricultural-Environmental Indicators (1989-91 to 2000). Statistics Analysis Service, Statistics Division, Food and Agriculture Organization of United Nations, Rome.

FAO. 2003b. *World agriculture: towards 2015/2030. An FAO perspective*, J. Bruinsma, ed. Rome, and London and New York, USA, Earthscan.

FAO. 2003c (continually updated). *Statistical databases.* (available at **http://www.fao.org**).

FAO. 2004. *Carbon sequestration in dryland soils.* World Soil Resources Reports No. 102. Rome.

FAO. 2005. *Integrated management for sustainable use of salt-affected soils.* FAO Soils Bulletin.

FAO. 2006. *World agriculture: towards 2030/2050.* Interim Report. Global Perspective Studies Unit. Food and Agriculture Organization of the United Nations. Rome.

FAO. 2007. *Conservation agriculture.* **http://www.fao.org/ag/ca**

FAO. 2008. Climate change, water and food security: a synthesis paper emanating from expert group meeting. 26-28 February, 2008. Food and Agriculture Organication of United Nations, Rome.

FAOSTAT. 2007. Statistical databases. Food and Agriculture Organization of United Nations, Rome. (available at http://faostat.fao.org/default.aspx).

Farshad, A. & Zinck, J.A. 1998. Traditional irrigation water harvesting and management in semiarid western Iran: a case study of the Hamadan region. *Wat. Int.*, 23: 146–154.

Feder, G., Murgai, R., & Quizon, J.B. 2004. Sending farmers back to school: the impact of farmer field schools in Indonesia. *Review of Agricultural Economics* 26:45-62.

Ffolliott, P., Dawson, J.O., Fisher, J.T., Moshe, I., Fulbright, T.E., Al Musa, A, Johnson, C. & Verburg, P. 2002. Dryland environments. *Ar. Land. News.*, 52.

FiBL. 2000. *Organic farming enhances soil fertility and biodiversity. Results from a 21 year field trial.* FiBL Dossier 1. Zurich, Switzerland, Research Institute of Organic Agriculture (FiBL).

Finnell, H.H. 1948a. The dust storms of 1948. *Sci. Am.*, 179: 7–11.

Finnell, H.H. 1948b. *Dust storms come from the poorer lands.* USDA Leaflet 260. Washington, DC, USDA.

Fowler, R. & Rockström, J. 2001. Conservation tillage for sustainable agriculture. An agrarian revolution gathers momentum in Africa. *Soil & Tillage Research* 61:93-107

Fresco, L.O. & Kroonenberg, S.B. 1992. Time and spatial scales in ecological sustainability. *Land Use Pol.*, 9(3): 155–168.

Fryrear, D.W. 1985. Wind erosion on arid croplands. *In: Science reviews, arid zone research*, pp. 31–48. Jodhpur, India, Scientific Publishers.

Furley, P.A. & Ratter, J.A. 1988. Soil resources and plant communities of the Central Brazilian cerrado and their development. *Journal Biogeography* 15:97-108.

Ghassemi, F.A., Jakeman, J. & Nix, H.A. 1995. *Salinization of land and water resources. Human causes, extent, management and case studies.* Wallingford, UK, CAB International.

Gisladottir, G. & Stocking, M. 2005. Land degradation control and its global environmental benefits. *Land Degradation and Development* 16:99-112.

Gras, N.S.B. 1946. *A history of agriculture in Europe and America.* New York, USA; F.S. Crofts & Co.

GRDC. 2008. An introduction to rainwater harvesting. Global Development Research Center, Kobe, Japan. (**http://www.gdrc.org/uem/water/rainwater/introduction.html**)

Haas, H.J., Willis, W.O. & Bond, J.J. 1974. *Summer fallow in the Western United States.* Conservation Research Report No. 17. Washington, DC, USDA Agricultural Research Service.

Han Siming, Si Juntung & Yang Chunfeng. 1988. Research on stubble mulch tillage on rainfed land. *In* P.W. Unger, T.V. Sneed, W.R. Jordan & R. Jensen, eds. *Challenges in dryland agriculture: a global perspective*, pp. 504–506. Proc. International Conference on Dryland Farming, Amarillo/Bushland, USA, 15–19 August 1988. USA, Texas Agricultural Experiment Station.

Hanks, R.J., Allen, L.H. & Gardner, H.R. 1971. Advection and evapotranspiration of wide-row sorghum in the central Great Plains. *Agron. J.*, 62: 520–527.

Hargreaves, M.W.M. 1957. *Dry farming in the Northern Great Plains: 1900–1925.* Cambridge, USA, Harvard University Press.

Harlan, J.R. 1976. Genetic resources in wild relatives of crops. *Crop Science* 16:329-333.

Hatibu, N. & Mahoo, H. 1999. Rainwater harvesting technologies for agricultural production: a case for Dodoma, Tanzania. *In* P.G. Kaumbaho & T.E. Simalenga, eds. *Conservation tillage with animal traction*, p. 161. Harare, ATNESA.

Hatibu, N. & Mahoo, H. 2000. *Rainwater harvesting for natural resources management: a planning guide for Tanzania.* Relma, Kenya (also available at **http://www.rainwaterharvesting.org**).

Hazell, P. 1998. *Why invest more in the sustainable development of less-favored lands?* IFPRI Report No. 20(2). Washington, DC, International Food Policy Research Institute.

Hegde, B.R. 1995. Dryland farming: past progress and future prospects. *In* R.P. Singh, ed. *Sustainable development of dryland agriculture in India*, pp. 7–12. Jodhpur, India, Scientific Publishers.

Hellig, G.K. 1999. *China food. Can China feed itself?* CD-ROM Version 1.1. Laxenburg, Austria, IIASA.

Hong Wei, Jian-Long Li & Tian-Gang Liang. 2005. Study on the estimation of precipitation resources for rainwater harvesting agriculture in semi-arid land of China. *Agric. Wat. Man.*, 71: 33-45.

Howell, T.A. 1990. Water use by crops and vegetation. *In* B.A. Stewart & D.R. Nielsen, eds. *Irrigation of agricultural crops*, pp. 391–434. Agronomy No. 30. Madison, USA, American Society of Agronomy, Crop Science Society of America and Soil Science Society of America.

Howell, T.A. 1998. Using the PET network to improve irrigation water management. *In* L.L. Triplett, ed. *The Great Plains symposium 1998: the Ogallala aquifer, determining the value of water*, pp. 38–45. Proc. 1998 Great Plains symposium, Lubbock, 10–12 March 1998. Overland Park, USA, Great Plains Foundation.

Hsu, H., Lohmar, G. & Gale, F. 2001. Surplus wheat production brings emphasis on quality. *In* H. Hsu & F. Gale, coordinators. *China: agriculture in transition*, pp. 17–25. Washington, DC, Economic Research Service, U.S. Department of Agriculture

Huanghe River Conservancy Commission. 1988. *Soil and water conservation in the Huanghe river valley.* Shanghai, China, Educational Publishing House.

Hurni, H. 1984. *Third progress report (year 1983).* Soil Conservation Research Project. Switzerland, University of Berne.

Hurni, H. 1993. Land degradation, famine, and land resource scenarios in Ethiopia. *In* D. Pimentel, ed. *World soil erosion and conservation*, pp. 27–61. Cambridge, Cambridge University Press.

IEA. 2005. International Energy Annual. (available at **http://www.eia.doe.gov/emeu/iea/contents.html**). Energy Information Administration, Washington, DC.

IFPRI. 1998. *Strategies for the development of fragile areas of sub-Saharan Africa.* IFPRI Report 20(2). Washington, DC.

IFPRI, 2001 Pilot Analysis of Global Ecosystems (PAGE) Agro-ecosystems, Washington, DC.

Institute for Agriculture and Trade Policy. 2006. Staying Home: How Ethanol will Change U.S. Exports. 2105

First Avenue South, Minneapolis, MN.

IPCC. 2007. *Climate Change 2007: Synthesis Rerort.* International Panel on Climate Change Report. Cambridge University Press, Cambridge, U.K.

IPTRID. 2001. *Case studies programme for technology and research in irrigation and drainage.* Knowledge Synthesis Report No. 4. Rome, IPTRID, FAO.

IWMI. 2003. Enhancing food security through small-scale innovations. Research update. Special issue. *Afr. Proj. Part.*, Oct./Nov.

IWMI. 2006. *Water for Food, Water for Life.* International Water Management Institute, Earthscan, London.

Jackson, I.J. 1989. *Climate, water and agriculture in the tropics.* Longman, Singapore.

Joel, A.H. 1937. *Soil conservation reconnaissance survey of the Southern Great Plains wind-erosion area.* Technical Bulletin No. 556. Washington, DC, USDA.

Johnson P.M., Mayrand, K. Paquin M. 2006. *Governing Global Desertification, Linking Environmental Degradation, Poverty and Participation* Ashgate Publishing Limited, Hampshire, England.

Jones, O.R. & Johnson, G.L. 1997. *Evaluation of a short-season - high-density production strategy for dryland sorghum.* Report No. 97-01. Bushland, USA, Conservation and Production Research Laboratory, USDA Agricultural Research Service.

Jones, O.R., Unger, P.W. & Fryrear, D.W. 1985. Agricultural technology and conservation in the Southern High Plains. *J. Soil Wat. Con.*, 40: 195–198.

Kätterer, T. & Andrén, O. 1999. Long-term agricultural field experiments in N. Europe: analysis of the influence of management on soil stocks using the ICBM model. *Agric., Ecosys. Env.*, 72: 165–179.

Kaumbutho, P.G. & Simalenga, T.E., eds. 1999. *Conservation tillage with animal traction.* Harare, ATNESA.

Kerr, J., in collaboration with Pangre, G. & Pangare, V.S. 2002.Watershed Development Projects in India. Research Report 127. International Food Policy Research Institute, Washington, D.C.

Kerr, J., Pangare, G., Pangare & Pangare, P.J. 2000. An evaluation of dryland watershed development in India. EPTD Discussion Paper 68. International Food Policy Research Institute, Washington, DC, USA.

Kerr, J. & Sanghi, N.K. 1992. *Indigenous soil and water conservation in India's semiarid tropics.* Gatekeeper Series No. 34. London, International Institute for Environment and Development.

Kirkham, M.B. (ed.). 1999. *Water Use in Crop Production.* The Haworth Press, Inc. New York.

Kirschke, D., Morgenroth, S. & Franke, C. 1999. How do human-induced factors influence soil erosion in developing countries? The role of poverty, population pressure and agricultural intensification. Paper presented at the International workshop assessing the impact of agricultural research on poverty alleviation, San José, 14–16 September 1999.

Koohafkan, A.P. 1996. Desertification, drought and their consequences. SD Dimensions. http://www.fao.org/waicent/faoinfo/sustdev/EPdirect/EPan0005.htm

Krieg, D.R. & Lascano, R.J. 1991. Sorghum. *In* B.A. Stewart & D.R. Nielsen, eds. *Irrigation of agricultural crops*, pp. 719–739. Madison, USA, American Society of Agronomy, Crop Science Society of America and Soil Science Society of America..

Kruska, R.L., Perry, B.D. & Reid, R.S. 1995. Recent progress in the development of decision support systems for improved animal health. *In: Integrated geographic information systems useful for a sustainable management of natural resources in Africa.* Proc. Africa GIS 1995 meeting, Abidjan.

LADA, 2008. *Guidelines For Land Use System Mapping.Technical Report # 8.* FAO, Rome.

Laflen, J.M., Moldenhauer, W.C. & Colvin, T.S. 1981. Conservation tillage and soil erosion on continuously row-cropped land. *In: Crop production with conservation tillage in the 80s*, pp. 121–133. ASAE Publication 7-81. St. Joseph, USA, American Society of Agricultural Engineers.

Lal, R. 1987. Managing the soils of sub-Saharan Africa. *Science*, 236: 1069–1076.

Lal, R. 1993. Soil erosion and conservation in West Africa. *In* D. Pimentel, ed. *Soil erosion and conservation*, pp. 1–25. Cambridge, Cambridge University Press.

Lal, R. 1997. Residue management, conservation tillage and soil restoration for mitigating greenhouse effect by CO_2-enrichment. *Soil Tillage Research* 43:81-107

Lal, R. 2002a. Carbon sequestration in dryland ecosystems in West Asia and North Africa. *Land Degradation and Development* 13:45-59.

Lal, R. 2002b. Soil carbon dynamics in cropland and rangeland. *Environmental Pollution* 116:352-362.

Lal, R. 2004. Soil carbon sequestration impacts on global climate change and food security. *Science* 304:1623-1627.

Lal, R. & Pierce, F.J. 1991. Soil management for sustainability. Soil and Water Conservation Society in cooperation with World Association Soil and Water Conservation and Soil Science Society of America, Ankeny, IA.

Lantican, M.A., Pingali, P.L. & Rajaram, S. 2003. Is research on marginal lands catching up? The case of unfavourable wheat growing environments. *Agricultural Economics* 29:353-361

Li Shengxiu & Xiao Ling. 1992. Distribution and management of drylands in the People's Republic of China. *Adv. Soil Sci.*, 18: 148–302.

Lockeretz, W., Shearer, G. & Kohl, D.H. 1981. Organic farming in the Corn Belt. *Science*, 211: 540–547.

Ma Shijun. 1988. Advances in mulch farming in China. *In* P.W. Unger, T.V. Sneed, W.R. Jordan & R. Jensen, eds. *Challenges in dryland agriculture: a global perspective*, pp. 510–511. Proc. International Conference on Dryland Farming, Amarillo/Bushland, USA, 15–19 August 1988. USA, Texas Agricultural Experiment Station.

Macartney, J.C., Northwood, P.J., Dagg, M. & Lawson, R. 1971. The effect of different cultivation techniques on soil moisture conservation and the establishment and yield of maize at Kongwa, Central Tanzania. *Trop. Agric. Trin.*, 48: 9–23.

Mann, T.L.J. 1991. Integration of crops and livestock. *In* V. Squires & P. Tow, eds. *Dryland farming: a systems approach*, pp. 102–118. South Melbourne, Australia, Sydney University Press.

Mathews, O.R. & Cole, J.S. 1938. Special dry-farming problems. *In: Soils and men*, pp. 679–692. Yearbook of Agriculture. Washington, DC, U.S. Department of Agriculture.

Mazzucato, V. & Niemeijer, D. 2000. The cultural economy of soil and water conservation: market principles and social networks in eastern Burkina Faso. *Dev. Change*, 31(4): 831–855.

McLaughlin, L. 1993. A case study in Gansu Province, China. *In* D. Pimentel, ed. *World soil erosion and conservation*, pp. 87–107. Cambridge, Cambridge University Press,.

Molden, D., Frenken, K., Barker, R., de Fraiture, C., Mati, B., Svendsen, M, Sadoff, C., & Finlayson, C.M. 2007. Trends in water and agricultural development. *In* D. Molden, ed. *Water for Food, Water for Life*. pp. 57-89. Earthscan, London and International Water Management Institute, Colombo.

Molden, D. & Oweis, T.Y. 2007. Pathways for increasing agricultural water productivity. *In* D. Molden, ed. *Water for Food, Water for Life*. pp. 279-310. Earthscan, London and International Water Management Institute, Colombo.

Morris, M.L., Belaid, A. & Byerlee, D. 1991. *Wheat and barley production in rainfed marginal environments of the developing world*. Part 1 of 1990–91 CIMMYT World wheat facts and trends. Mexico, CIMMYT.

Mortimore, M. 1998. *Roots in the African dust: sustaining the Sub-Saharan drylands.* Cambridge University Press, Cambridge, U.K.

Mortimore, M. & Adams, W.M. 1999. *Working the Sahel: environment and society in northern Nigeria.* London: Routledge

Mrabet, R. 2002. Stratification of soil aggregation and organic matter under conservation tillage systems in Africa. *Soil & Tillage Research* 66:119-128.

Musick, J.T. & Porter, K.B. 1990. Wheat. *In* B.A. Stewart & D.R. Nielsen, eds. *Irrigation of agricultural crops.* Agronomy 30. Madison, USA, American Society of Agronomy, Crop Science Society of America and Soil Science Society of America.

Musick, J.T., Jones, O.R., Stewart, B.A. & Dusek, D.A. 1994. Water-yield relationships for irrigated and dryland wheat in the U.S. Southern Plains. *Agron. J.*, 86: 980–986.

Nasr, M. 1999. Assessing desertification and water harvesting in the Middle East and North Africa. *Policy Implication Discussion Paper* No. 10. Center for Development Research (ZEF), Bonn, Germany.

National Weather Service, 2004. What is meant by the term drought? U.S. Department of Commerce, Washington, DC. (available at **http://www.wrh.noaa.gov/fgz/science/drought.php**)

Njihia, C.M. 1979. The effect of tied ridges, stover mulch, and farmyard manure on water conservation in a medium potential area, Katumani, Kenya. *In* R. Lal, ed. *Soil tillage and crop production*, pp. 295–302. Ibadan, Nigeria, IITA.

Oberthür, S, & Ott, H.E. 1999. *The Kyoto Protocol: intenational climate policy for the 21ˢᵗ century.* Springer, Berlin, Heidelberg, New York.

Oldeman, L.R., Hakkeling, R.T.A. & Sombroek, W.A. 1991. *Map of the status of human-induced soil degradation. An explanatory note.* Wageningen, The Netherlands, International Soil Reference and Information Centre and United Nations Environment Programme.

Oram, P. 1980. What are the world resources and constraints for dryland agriculture? *In: Proceedings international congress dryland farming*, pp. 17–78. Adelaide, Australia, South Australia Department of Agriculture.

Ouessar, M., Sghaier, M. & Fetoui, M. 2002. *A comparison of the traditional and contemporary water management systems in the arid regions of Tunisia.* UNU–UNESCO–ICARDA Joint International Workshop on Sustainable Management of Marginal Drylands, Application of Indigenous Knowledge for Coastal Drylands, Alexandria, Egypt.

Oweis, T. 1996. *The lost resource. Caravan 2.* Aleppo, Syrian Arab Republic, ICARDA.

Oweis, T. 1997. *Supplemental irrigation – a highly efficient water-use practice.* ICARDA-037/1000. pp. 1–16.

Oweis, T. 2000. *Management of scarce water resources in agriculture.* Proc. New Approaches to Water Management in Central Asia, workshop. United Nations University, and Aleppo, Syrian Arab Republic, ICARDA.

Oweis, T., Hachum, A. & Kijne, J. 1999. *Case studies on water conservation. Water harvesting and supplementary irrigation for improved water-use efficiency in dry areas.* SWIM paper No. 7. Colombo, IWMI.

Oweis, T., Prinz, D. & Hachum, A. 2001. *Water harvesting: indigenous knowledge for the future of the drier environments.* Aleppo, Syrian Arab Republic, ICARDA.

Oweis, T., Hachum, A., & Pala, M. 2004. Lentil production under supplemental irrigation in a Mediterranean environment. *Agricultural Water Management* 68:251-256

Pagiola, S. 2005. Can payments for environmental services help reduce poverty? An exploration of the issues and the evidence to date from Latin America. *World Development* 33:237.

Pala, M. & Mazid, A. 1992. On-farm assessment of improved crop production practices in northwest Syria: chickpea. *Exp. Agric.*, 28: 175–184.

Pender, J. 1999. *Rural population growth, agricultural change and natural resource management in developing countries: a review of hypotheses and some evidence from Honduras.* EPTD Discussion Paper No. 48. Washington, DC, IFPRI.

People's Daily Online. 2006. China halts expansion of ethanol industry. http://english.peopledaily.com.cn/200612/20/eng20061220_334361.html

Perrier, E.R. & Salkini, A.B. 1991. Introduction. *In* E.R. Perrier & A.B. Salkini, eds. *Supplemental irrigation in the Near East and North Africa*, pp. 1–14. Dordrecht, The Netherlands, Kluwer Academic Publishers.

Petersen, C., Drinkwater, L.E. & Wagoner, P. 2000. *The Rodale Institute's farming systems trial. The first 15 years.* USA, Rodale Institute.

Pieri, C. 1995. Long-term soil management experiments in semiarid francophone Africa. *In* R. Lal & B.A. Stewart, eds. *Soil management: experimental basis for sustainability and environmental quality*, pp. 225–266. Boca Raton, USA, CRC Press.

Pingali, P.L. & Pandey, S. 2000. Meeting world maize needs: technological opportunities and priorities for the public sector. *In* P.L. Pingali, ed. *CIMMYT 1999/2000 World maize facts and trends. Meeting world maize needs: technological opportunities and priorities for the public sector*, pp. 1–3. Mexico, CIMMYT.

Pingali, P.L. & Rajaram, S. 1999. Global wheat research in a changing world: options for sustaining growth in wheat productivity. *In* P.L. Pingali, ed. *CIMMYT 1999/2000 World maize facts and trends. Meeting world maize needs: technological opportunities and priorities for the public sector*. Mexico, CIMMYT.

Ponce, V.M., 1995. Management of droughts and floods in the semiarid Brazilian Northeast—the case for conservation. *Soil Water Conserv. J.* 50:422-431.

Post, W.M. & Kwon, K.C. 2000. Soil carbon sequestration and land-use change: processes and potential. *Global Change Biology* 6:317-327.

Postel, S. 1999. *Pillar of sand: can the irrigation miracle last?* New York, USA, W.W. Norton & Company.

Powell, J.M., Pearson, R. & Hiernaux, P. 2004. Crop-livestock interactions in the west African highlands. *Agronomy Journal* 96:469-483.

Pretty, J. & Ball, A. 2001. *Agricultural influences on carbon emissions and sequestration: a review of evidence and the emerging trading options.* CES Occasional Paper 2001-3. Colchester, UK, University of Essex.

Pretty, J. & Koohafkan, A.P. 2002. Land and Agriculture, from UNCED Rio de Janeiro 1992 to WWSD, Johannessburg 2002. Food and Agriculture Organization of United Nations, FAO, Rome.

Prinz, D. & Wolfer, S. 1998. *Opportunities to ease water scarcity (water conservation techniques and approaches).* Proc. international conference on world water resources at the beginning of the 21st century, 3–6 June 1998, Paris. UNESCO-IHP, Paris (also available at **http://www.ubka.uni-karlsruhe.de**).

Radder, G.D., Belgaumi, M.I. & Itnal, C.J. 1995. Water harvesting procedures for dryland areas. *In* R.P. Singh, ed. *Sustainable development of dryland agriculture in India*, pp. 195–205. Jodhpur, India, Scientific Publishers.

Rasmussen, K. 1999. Land degradation in the Sahel–Sudan: the conceptual basis. *Geog. Tids.*, 2 (Special Issue): 151–159.

Rasmussen, P.E., Goulding, K.W.T., Brown, J.R., Grace, P.R., Janzen, H.H. & Körschens, M. 1998. Long term agroecosystem experiments: assessing agricultural sustainability and global change. *Science*, 282: 893–896.

Rathore, M.S. 2005. Groundwater exploration and augmentation efforts in Rajasthan – a review. Institute of Development Studies, Jaipur, India.

Reganold, J.P., Elliott, L.F. & Unger, Y.L. 1987. Long-term effects of organic and conventional farming on soil erosion. *Nature*, 330: 370–372.

Reganold, J.P., Palmer, A.S., Lockhart, J.C. & Macgregor, A.N. 1993. Soil quality and financial performance of biodynamic and conventional farms in New Zealand. *Science*, 260: 344–349.

Reicosky, D.C., Dugs, W.A. & Torbert, H.A. 1997. Tillage-induced soil carbon dioxide loss from different cropping systems. *Soil Tillage Res.*, 41: 105–118.

Reicosky, D.C., Kemper, W.D., Langdale, G.W., Douglas, C.L. & Rasmussen, P.E. 1995. Soil organic matter changes resulting from tillage and biomass production. *J. Soil Wat. Con*, 50: 253–261.

Reij, C., Mulder, P. & Begemann, L. 1988. *Water harvesting for plant production. World Bank Technical Paper* No. 91.

Reij, C. & Waters-Bayer, A. 2001. *Farmer Innovation in Africa: A Source of Inspiration for Agricultural Development.* Earthscan, London.

Rhoads, F.M. & Bennett, J.M. 1991. Corn. *In* B.A. Stewart & D.R. Nielsen, eds. *Irrigation of agricultural crops*, pp. 569–596. Agronomy 30. Madison, USA, American Society of Agronomy, Crop Science Society of America and Soil Science Society of America.

Ritchie, J.T. 1983. Efficient water use in crop production: discussion on the generality of relations between biomass production and evapotranspiration. *In* H.M. Taylor, W.R. Jordan & T.R. Sinclair. *Limitations to efficient water use in crop production*, pp. 29–44. Madison, USA, American Society of Agronomy, Crop Science Society of America and Soil Science Society of America.

Ringler, C., Rosegrant, M., Cai, X., & Cline. S. 2003. Auswirkungen der zunehmenden Wasserverknappung auf die globale und regionale Nahrungsmittelprodukion. *Zeitschrift für angewandte Umweltforschung (ZAU)* 15/16 (3-5): 604-619.

Rockström, J. 2000. Water resources management in smallholder farms in eastern and southern Africa: an overview. *Phys. Chem. Earth (B)* 25:275-283.

Rockwood, R.W. & Lal, R. 1974. Mulch tillage: a technique for soil and water conservation in the tropics. *Span*, 17: 72–79.

Rosegrant, M.W. & Ringler, C. 1999. *World Water Vision scenarios: consequences for food supply, demand, trade, and food security: results from the IMPACT implementation of the World Water Vision scenarios.* Washington, DC, IFPRI.

Rosegrant, M.W., Cai, X., Cline, S. & Nakagawa, N. 2002. *The role of rainfed agriculture in the future of global food production.* EPTD Discussion Paper No. 90. Washington, DC, EPTD, IFPRI.

Ruthenberg, H. 1980. *Farming systems in the tropics.* Third edition. Oxford, Clarendon Press.

Sanchez, P.D. 1997. Changing tropical soil fertility paradigms: from Brazil to Africa and back. *In* A.C. Moniz, *et al.* (eds.) pp. 19-28. *Plant–Soil Interactions at Low pH*, Brazilian Society of Soil Science, Piracicaba, SP,

Sandford, S. 1988. Integrated cropping-livestock systems for dryland farming in Africa. *In* P.W. Unger, T.V. Sneed, W.R. Jordan & R. Jensen, eds. *Challenges in dryland agriculture: a global perspective*, pp. 861–872. Proc. International Conference on Dryland Farming, Amarillo/Bushland, USA, 15–19 August 1988. USA, Texas Agricultural Experiment Station.

Sandor, J.A. & Eash, N.S. 1995. Ancient agricultural soils in the Andes of southern Peru. *Soil Sci. Soc. Am. J.*, 59: 170–179.

Saxena, N.C. 2001. How have the poor done? Mid-term review of India's ninth five-year plan. Natural Resource Perspectives. Overseas Development Institute. London.

Schlesinger, W.H. 2000. Carbon sequestration in soils: some cautions amidst optimism. *Agriculture, Ecosystems and Environment* 82:121-127.

Scoones, I., Reij, C. & Toulmin, C., eds. 1996. *Sustaining the soil: indigenous soil and water conservation in Africa*. London, Earthscan Publications Ltd.

Seckler, D., Amarasinghe, U., Molden, D., de Silva, R. & Barker, R. 1998. *World water demand and supply, 1990 to 2025: scenarios and issues*. Research Report 19. Colombo, IWMI.

Shiklomanov, I.A. 1999. *World water resources and water use: present assessment and outlook for 2025*. St. Petersburg, Russia, State Hydrological Institute.

Siadat, H. 1991. Potential of supplemental irrigation in Iran. *In* E.R. Perrier & A.B. Salkini, eds. *Supplemental irrigation in the Near East and North Africa*, pp. 367–388. Dordrecht, The Netherlands, Kluwer Academic Publishers.

Simpson, B. & Owens, M. 2002. Farmer field schools and the future of agricultural extension in Africa. SDdimensions, Sustainable Development Department, Food and Agriculture Organization of the United Nations, Rome.

Singh, K.P. & Ocompo, B. 1997. Exploitation of wild Cicer species for yield improvement in chickpea. *Theoretical and Applied Genetics* 95:418-423.

Singh, R.P. 1995. Dryland agricultural research in India – a historical perspective. *In* R.P. Singh, ed. *Sustainable development of dryland agriculture in India*, pp. 1–6. Jodhpur, India, Scientific Publishers.

Smika, D.E. 1976. Seed zone soil water conditions with mechanical tillage in the semiarid Central Great Plains. *In: Proceedings of seventh International Soil Tillage Research Organization conference*. Uppsala, Sweden, College of Agriculture.

Smith, K.A. 1999. After the Kyoto protocol: can scientists make a useful contribution? *Soil Use Man.*, 15: 71–75.

Smith, P.D. & Critchley, W.R.S. 1983. The potential of run-off harvesting for crop production and range rehabilitation in semi-arid Baringo. *In: Soil and water conservation in Kenya. Proc. Second National Workshop*, March 1982, Nairobi. Occ. Paper 42. Nairobi, University of Nairobi.

Smith, P., Powlson, D.S. Glendenning, M.J. & Smith, J.U. 1998. Preliminary estimates of the potential for carbon mitigation in European soils through no-till farming. *Glob. Change Biol.*, 4: 679–685.

Squires, V.R. 1991. A systems approach to agriculture. *In* V. Squires & P. Tow, eds. *Dryland farming: a systems approach*, pp. 3–15. Sydney, Australia, Sydney University Press.

Srivastava, J.P., Tamboli, P.M., English, J.C., Lal, R. & Stewart, B.A. 1993. *Conserving soil moisture and fertility in the warm seasonally dry tropics*. World Bank Technical Paper No. 221. Washington, DC, World Bank.

Steiner, J.L., Day, J.C, Papendick, R.I., Meyer, R.E. & Bertrand, A.R. 1988. Improving and sustaining productivity in dryland regions of developing countries. *Adv. Agron.*, 8: 79–122.

Stern, N. 2007, *The Economics of Climate Change*. Cambridge University Press, Cambridge, U.K.

Stern, N. 2006. *The Stern Review on the Economics of Climate Change*. Cambridge University Press, Cambridge, UK.

Stewart, B.A. & Burnett, E. 1987. Water conservation technology in rainfed and dryland agriculture. *In* W.R. Jordan, ed. *Water and water policy in world food supplies*, pp. 355–359. USA, Texas A&M University Press.

Stewart, B.A. & Koohafkan A.P. 2004. Dryland Agriculture, Long neglected but of worldwide importance. In Challenges and Strategies for Dryland Agriculture. *CSSA Special Publication* no. 32, pp. 49-60. Madison,

USA, American Society of Agronomy, Crop Science Society of America and Soil Science Society of America.

Stewart, B.A., Koohafkan, A.P. & Ramamoorthy, K. 2006. Dryland Agriculture defined and its importance to the World. In Dryland Agriculture, Agronomy Monograph No.23, Second Edition pp. 1-26. Madison, USA, American Society of Agronomy, Crop Science Society of America and Soil Science Society of America.

Stewart, B.A. & Robinson, C.A. 1997. Are agroecosystems sustainable in semiarid regions? *Adv. Agron.*, 60: 191–228.

Stewart, B.A. & Steiner, J.L. 1990. Water-use efficiency. *In* R.P. Singh, J.F. Parr & Stewart, B.A. eds. *Dryland agriculture: strategies for sustainability*, pp. 151–173. New York, USA, Springer Verlag.

Stewart, B.A., Jones, O.R. & Unger, P.W. 1993. Moisture management in semiarid temperate regions. *In* J.P. Srivastava & H. Alderman, eds. *Agriculture and environmental challenges*, pp. 67–80. Proc. Thirteenth Agricultural Sector Symposium. Washington, DC, World Bank.

Stewart, B.A. Lal, R. & El-Swaify, S.A. 1991. Sustaining the resource base of an expanding world agriculture. *In* R. Lal & F.J. Pierce, eds. *Soil management for sustainability*, pp. 125–144. Ankeny, USA, Soil and Water Conservation Society.

Stoskopf. N.C. 1985. *Cereal Grain Crops.* Reston Publishing Company, Reston, VA.

Studer, C. & Erskine, W. 1999. *Integrating germplasm improvement and agricultural management to achieve more efficient water use in dry area crop production.* Paper presented at the International Conference on Water Resources Conservation and Management in Dry Areas, 3–6 December 1999. Amman.

Suleman, S., Wood, K., Shad, B. & Murray, L. 1995. Development of a rainwater harvesting system for increasing soil moisture in arid rangelands of Pakistan. *J. Arid Env.*, 31: 471–481.

Sung-Chiao. 1981. *Desert lands of China. I. The sandy deserts and the Gobi: a preliminary study of their origin and evolution.* ICASALS Publ. No. 81-1. Lubbock, USA, International Center for Arid and Semiarid Land Studies, Texas Tech. University.

Svendsen, M. & Turral, H. 2007. Reinventing irrigation. In D. Molden, ed. *Water for Food, Water for Life.* pp. 353-394. Earthscan, London and International Water Management Institute, Colombo.

Tebrügge, F. 2000. *No-tillage visions - protection of soil, water and climate.* Giessen, Germany, Justus-Liebig University.

Tian Houmo. 1985. The sharp increase of population and soil erosion in the Loess Plateau. *Bull. Soil Wat. Con.*, 3: 7–11.

Tiessen, H., Cuevas, B. & Chacon, P. 1994. The role of soil organic matter in sustaining soil fertility. *Nature*, 371: 783–785.

Tiffen, M., Mortimore, M. & Gichuki, F. 1994. *More people, less erosion: environmental recovery in Kenya.* Chichester, UK, John Wiley & Sons.

Tilman, D. 1998. The greening of the green revolution. *Nature*, 396: 211–212.

Tobler, W., Deichmann, U., Gottsegen, J. & Maloy, K. 1995. *The global demography report.* Technical Report TR-95-6, NCGIA. Santa Barbara, USA, Geography Department, University of California. 75 pp + diskette.

Tow, P.G. 1991. Factors in the development and classification of dryland farming systems. *In* V. Squires & P. Tow, eds. *Dryland farming: a systems approach*, pp. 24–31. Sydney, Australia, Sydney University Press.

Tow, P.G. & Schultz, J.E. 1991. Crop and crop-pasture sequences. *In* V. Squires & P. Tow, eds. *Dryland farming: a systems approach*, pp. 55–75. Sydney, Australia, Sydney University Press.

Tubiello, F. & Fischer, G. 2007. *Reducing climate change impacts on agriculture: Global and regional effects of mitigation, 2000-2080.* Technol. Forecasting Soc. Change, 74:1030-1056.

UNCCD. 2000. *An introduction to the United Nations Convention to combat desertification* (available at **http://www.unccd.int**).

UNDP. 2006. The United Nations World Water Development Report 2. UNESCO, Paris and Berghahn Books, New York.

UNEP. 1982. Rain and storm water harvesting in rural areas. United Nations Environment Programme. Tycooly International Publishing Ltd., Dublin.

UNEP. 2000. *Sourcebook of alternative technologies for freshwater augmentation in West Asia* (available at **http://www.unep.or.jp**).

UNESCO. 1977. World map of desertification. A/Conference 74/2. Rome, FAO.

Unger, P.W. 1978. Straw-mulch rate effect on soil water storage and sorghum yield. *Soil Sci. Soc. Am. J.*, 42: 486–491.

Unger, P.W. & Baumhardt, R.L. 1999. Factors related to dryland grain sorghum yield increases. *Agron. J.*, 91: 870–875.

Unger, P.W. & Jones, O.R. 1981. Effect of soil water content and a growing season straw mulch on grain sorghum. *Soil Sci. Soc. Am. J.*, 45: 129–134.

Unger, P.W. & Parker, J.J. 1976. Evaporation reduction from soil with wheat, sorghum and cotton residues. *Soil Sci. Soc. Am. J.*, 40: 938–942.

UNDP/UNSO, 1997. *Aridity zones and drylands populations.*

UNFPA, 2008. *World Population Prospects: The 2006 Revision* and *World Urbanization Prospects: The 2005 Revision,* **http://esa.un.org/unpp.**

USDA. 1997. *1997 census of agriculture.* Washington, DC, National Agricultural Statistics Service, USDA.

Van de Fliert, E. 1993. Integrated pest management: Farmer field schools generate sustainable practices: A case study in central Java evaluating IPM training. WU Papers 93-3. Published doctoral dissertation. Wageningen: Agricultural University.

Varisco, D.M. 1991. The future of terrace farming in Yemen: a development dilemma. Agriculture and Human Values 8:166-172.

Venkataraman, A. 1999. Helping farmers to help themselves irrigate Balochistan. World Bank Group (also available at http://wbln0018.worldbank.org).

Viets, F.G., Jr. 1962. Fertilizers and the efficient use of water. *Adv. Agron.*, 14: 223–264.

Wander, M., Bidart, M. & Aref, S. 1998. Tillage experiments on depth distribution of total and particulate organic matter in 3 Illinois soils. *Soil Sci. Soc. Am. J.*, 62: 1704–1711.

Wen Dazhong. 1993. Soil erosion and conservation in China. *In* D. Pimentel, ed. *World soil erosion and conservation,* pp. 63–85. Cambridge, Cambridge University Press.

White, R.P. & Nackoney, J. 2003. *Drylands, people, and ecosystem goods and services: a web-based geospatial analysis* (available at **http://pubs.wri.org**).

White, R.P., Tunstall, D. & Henninger, N. 2002. An ecosystem approach to drylands: building support for new development policies. Information Policy Brief No. 1, World Resources Institute, Washington, DC.

WHO, 2008. Global and regional food consumption patterns and trends. World Health Organization, (available at **http://www.who.int/nutrition/topics/3_food consumption/en/index/html**).

WMO. 1990. *Glossary of terms used in agrometeorology.* CAGM No. 40. WMO/TD-No. 391. Geneva.

WOCAT. 2007. Where the land is greener. (available at **http://www.wocat.net/overviewbook.asp**). World Overview of Conservation Approaches and Technologies, Berne, Switzerland.

Wood, S., Sebastian, K. & Scherr, S.J. 2000. *Agroecosystems. Pilot analysis of global ecosystems.* Washington, DC, IFPRI and World Resources Institute.

World Bank. 2000. *World development indicators.* Washington, DC.

World Bank, 2006. Reengaging in Agricultural Water Management: Challenges and Options. Washington, DC.

Worldwatch Institute. 2006. Biofuels for Transportation: Global Potential and Implications for Sustainable Agriculture and Energy in the 21st Century. Worldwatch Institute, Washington, DC.

WRI. 2005. More water, more wealth in Darewadi Village. World Resources Institute, Washington, DC. (available at **http://www.wri.org/publication/content/7589**) .

Xi Chengfan. 1961. Soil moisture and conservation of water for resisting drought. *Soils,* 4: 27–29.

Yang Zhenhuai. 1986. Solidarity urged for opening up new prospects in conservation practices. *Soil Wat. Con. Chin.,* 8: 7–12.

Zhang Jinhui. 1987. The cost-benefit analysis of the comprehensive controls in Wangmaogou small watershed. *Soil Wat. Con. Chin.,* 7: 40–43.

Definitions of Drylands and Dryland farming

Dryland farming antedates history, but Hegde (1995) stated that the use of the term in its present form and meaning probably began in Utah, the United States of America, around 1863 and credits John A. Widstoe as being the pioneer of dryland farming research. According to Hegde, Widstoe defined dry farming as "the profitable production of useful crops without irrigation on lands that receive rainfall of less than 500 mm annually", and said that the definition could be extended to include areas receiving up to 750 mm annual rainfall where its distribution was unfavourable. It is important to note that this definition is limited to land where crops are grown and does not include the management of other lands in dryland regions.

Mathews and Cole (1938) broadened the definition by suggesting that dry farming is concerned with all phases of land use under semi-arid conditions. Not only the question how to farm but also how much to farm and whether to farm or not must be taken into consideration. These questions raised by Mathews and Cole are critical regarding whether or not dryland agriculture can be sustained in the long term. However, they have not received the attention they deserve. Hargreaves (1957) defined dry farming as agriculture without irrigation in regions of scanty precipitation.

Oram (1980) made the distinction between rainfed farming and dryland agriculture explicit. He defined dryland agriculture as husbandry under conditions of moderate to severe moisture stress during a substantial part of the year, which requires special cultural techniques and crops and farming systems adapted for successful and stable agricultural production. Such conditions generally occur in regions classified as semi-arid or arid. Pastoral systems are an important part of dryland agriculture, and constitute the sole form of agricultural use in some areas, particularly in arid regions. Stewart and Burnett (1987) added that dryland farming emphasizes water conservation in all practices throughout the year. The Australian Centre for International Agricultural Research (2002) considers dryland cropping to occur in areas where the average water supply to the crop limits yield to less than 40 percent of full (not water-limited) potential. On this basis, it estimates that one-quarter of the world's cereal production is from dryland agriculture.

FAO (2000a) defined drylands as those regions classified climatically as arid, semi-arid, or dry subhumid, based on the length of the growing period for annual crops. The growing period begins when monthly precipitation exceeds half of the monthly potential evapotranspiration. The regions where the monthly rainfall never exceeds half of the potential evapotranspiration have zero growing days and are not included in the drylands. They are classified as hyperarid areas with no agricultural potential.

Arid regions have 1–59 growing days, semi-arid have 60–119 growing days, and dry subhumid regions have 120–179 growing days. Together, these regions make up 45 percent of the world's land area: 7 percent arid, 20 percent semi-arid, and 18 percent dry subhumid. The distribution of these areas among the different regions of the world is presented in Figure 1 and in Table 1 in the main text of this book. The table does not include the hyperarid lands, which make up an additional 19 percent of the world's land area.

The classification system developed by FAO based on agro-ecological factors generally works well for assessing the potential of an area for growing crops, but there are notable exceptions. For example, in much of the Great Plains of the United States of America, one of the largest dryland cropping regions in the world, there is not a single month of the year when average precipitation exceeds half of the reference evapotranspiration. Based on the above classification system, this area is classified as hyperarid and considered non-agricultural. However, crops can be grown in this region because management practices have been developed that allow the accumulation of 100–200 mm of plant-available water in the soil during fallow periods to supplement precipitation received during the growing season. These cropping systems have a cropping intensity (the proportion of land planted to each crop during a year) of less than one, meaning that a crop is not harvested every year. Examples of such cropping systems are: wheat–fallow, resulting in one crop every two years; wheat–sorghum–fallow, resulting in two crops every three years; and wheat–sorghum–sorghum–fallow, resulting in three crops every four years. The average annual precipitation for the region where dryland farming is practised ranges from about 400 to 600 mm, but the amount for any given year for a specific location varies from about 50 percent of average annual amount to about 200 percent. The variation in yields is even greater, ranging from about 0 to 3 times the average yield. Drought conditions occur every year, but their extent and severity vary greatly.

Dryland areas such as the Great Plains may be better characterized by a climate aridity index. One such index proposed by the United Nations Conference on Desertification (UNESCO, 1977) defines bioclimatic zones by dividing the annual precipitation (P) by the annual potential evapotranspiration (PET). Climate zones are defined as hyperarid (P/PET < 0.03), arid (0.03 < P/PET < 0.20), semi-arid (0.20 < P/PET <0.50), and subhumid (0.50 < P/PET < 0.75). This aridity index classifies the Great Plains as a semi-arid region where dryland farming is widely practised, whereas the growing-period system showed this large region as hyperarid. There is probably no classification system that can be universally applied and each area should be considered on its own merits. National and regional classifications that reflect local characteristics have also been developed to inform decision-making processes.

Bowden (1979) divided semi-arid lands into two broad physical-climatic regions: the tropical semi-arid territory close to the equator; and the steppe located in the mid-latitudes. Each region has peculiarities of climate, settlement and resource development that can vary as much internally as externally. In contrast to the frost-free, semi-arid tropics, the semi-arid mid-latitude lands are characterized by definite warm and cold seasons. Bowden used annual precipitation between 250 and 550 mm to define the steppe. In these conditions, crop production is marginal and the growing season is often shortened by unseasonal frost. The summer precipitation is mainly evaporated or transpired during the warm season, but winter rainfall and snow usually contribute to stored water in the soil profile. The dominant vegetation under native conditions was grass, and ploughing of grassland was difficult without strong draught animals and the steel plough. Therefore, almost none of the world's steppe land was farmed extensively 150 years ago, except beside streams where shallow groundwater was available. Mechanization has led to greatly expanded agricultural production in the semi-arid areas of North America and Eurasia.

Development of Dryland farming in various regions

Dryland farming is practised in various parts of the world. The specific practices vary because of differences in local conditions, both physical and social. Development histories of some of the larger dryland regions of the world are presented below.

AUSTRALIA

Much of Australia's agricultural income comes from the production of food and fibre on dryland farms (Squires, 1991). In Australia, dryland-farming systems combine crops, pastures and fallow periods for the purpose of making efficient use of the limited water. Moisture is usually the deciding factor in the success of cereal cropping. Fallowing can be an important strategy to store and conserve water for the establishment and maturation of the crop. Winter crop production often depends on fallowing over the summer.

Australian dryland-farming systems have distinctive characteristics (Tow, 1991). These reflect: the importance of livestock products, particularly sheep, and wheat as export commodities; the availability of pasture legumes for incorporation into crop rotations; and the need for efficiency imposed by farm size. However, in general, Australian dryland-farming systems have similarities to three major world farming systems: Mediterranean agriculture; mixed farming of Western Europe and eastern North America; and large-scale grain production.

Mediterranean agriculture has had a large influence on the agriculture of southern Australia because the the of the Mediterranean-type climate and because of the importance of crop, pasture, weed and livestock species from the Mediterranean basin. Mixed farming areas of the United States of America featuring large farm size and dependence on mechanization are similar to Australian cereal producing farms (Tow, 1991). One similarity is the initial abundant supply of new fertile land and the relative scarcity of labour. Large-scale cereal production is common in parts of Argentina, Australia, Canada, the former Soviet Union and the United States of America, and their early successes came mainly through the exploitation of large areas of initially fertile lands brought under cultivation. Australian examples are the black, grey and brown clays (mainly Vertisols) of northern New South Wales and southern and central Queensland. Because of the initial high fertility, inputs of fertilizers were low for many years. However, these systems were not sustainable because of the lack of flexibility, risk of soil erosion, buildup of disease organisms, and periodic low prices (Tow, 1991). Australian farmers have had some success in finding alternative crop and pasture species and in overcoming erosion problems by maintaining crop residues on the soil surface through conservation tillage (Tow and Schultz, 1991; Cornish and Pratley, 1991).

CHINA

Arid and semi-arid lands in China cover 52 percent of the country – 31 and 21 percent, respectively (Li Shengxiu and Xiao Ling, 1992). These lands are located between 30 and 50

°N, from the warm temperate belt in the south to the temperate zone in the north. Rainfall is very variable; annual totals average 300–500 mm, and increase gradually from northwest to southeast. Rainfall from June to September accounts for 70–80 percent of the annual total.

Li Shengxiu and Xiao Ling (1992) extensively reviewed the extent, characteristics, and management of the drylands in China. Management has focused on transforming desert and desertified lands, controlling erosion, and making efficient use of the precipitation. Although the drylands make up more than half of the land area, they account for 30 percent of the arable land. The development of agriculture in these vast areas is extremely important to the nation.

The basic tillage principle for conserving soil and water in China is to increase soil-surface roughness (Li Shengxiu and Xiao Ling, 1992). This is done mainly by building low earthen banks between fields, making ridges and furrows, or digging ditches in fields. Examples include: contour ploughing, contour planting, digging pits for seeding, contour plough furrows, and cultivation in pits or furrows. Another important principle used in designing soil- and water-conserving cropping systems is to increase plant cover. Narrow crop rows, intercropping and interplanting are widely used for this purpose. These practices increase the density of the crop canopy, which reduces raindrop impact on the soil surface and surface sealing, maintains soil permeability, and reduces or eliminates runoff and erosion.

Fallowing has also been considered an important practice for restoring soil water and fertility in China. During the fallow period, weeds are controlled by cultivation. The fallow period differs depending on precipitation and soil fertility. The effect of a one-year fallow period on the subsequent crop production will last for at least 3 years.

Ploughing to a depth of about 20–30 cm has been used widely in dryland farming as an effective method for storing precipitation. Deep ploughing is generally carried out in the summer, but not in the spring. Summer ploughing keeps the soil loose during the rainy season of July–September, enhances water penetration, and reduces runoff. Spring ploughing increases water loss from the soil because this season is often dry (Li Shengxiu and Xiao Ling, 1992).

As in India, dust mulching is commonly used in China for conserving soil water, and research has shown it to be beneficial (Xi Chengfan, 1961). This practice is discouraged in North America. Stubble mulching has been important for controlling wind erosion on dryland fields in the United States of America following the Dust Bowl. It is a minimum tillage practice that aims to leave substantial amounts of crop residues on the soil surface to reduce evaporation. A study of this practice in China (Han Siming, Si Juntung and Yang Chunfeng, 1988) showed that water conservation from this practice was poor and that mulching after sowing was unfavourable to wheat. Wheat seedlings often became yellow and had more sharp and dry leaves when fertilizers were not applied. There was also a shortage of crop residues and this hampered crop residue management on the soil surface. Another major problem was the low soil temperature that had negative effects on the mineralization of soil nutrients and enhanced weed growth and plant diseases.

The use of plastic mulches in China has expanded rapidly. Ma Shijun (1988) reported that there were more than 1.3 million ha of plastic mulch in use, mainly in the northern provinces of Liaoning, Shanxi, Shaanxi and Shandong, and in the Xinjiang Autonomous Region. Plastic-film mulch was used on about half of the cotton and peanut fields and accounted for more than three-quarters of the fields using plastic mulch in 1985. The use of plastic mulches in China has continued to increase at a rapid rate. However, the driving factor for this practice has been the desire to increase soil temperature rather than reduce soil-water evaporation.

ETHIOPIA

Ethiopia is facing a tremendous challenge in meeting the food needs of a rapidly growing population. Famines have occurred frequently in Ethiopia, with the 1984–85 famine affecting almost ten million people (Hurni, 1993). Both irrigated and dryland cropping areas will have to be developed or improved in the future. These tasks will not be easy, inexpensive or swift.

There were 194 000 ha of irrigated land in Ethiopia in 1994, representing about 3.2 percent of the cultivated area (FAO, 1995). Ancient irrigation systems have a long history in Ethiopia, with modern systems dating from the 1960s. The first large irrigated farms, established by private investors, were located in the middle Awash Valley, where there are large sugar estates, and fruit and cotton farms. These farms became the responsibility of the Ministry of State Farms with the 1975 rural land proclamation. There are small-scale, medium-scale and large-scale irrigation systems in Ethiopia (FAO, 1995). Small-scale projects are smallholder projects for a single peasant association up to 200 ha in size. Medium-scale systems are between 200 and 3 000 ha, extending beyond one peasant association. Large-scale systems are centrally managed state farms for commercial production and cover 3 000 ha or more.

In 1988, the costs of developing large-scale schemes were US$18 000–25 000/ha, without including water storage. Development costs of medium-scale schemes were US$10 000–15 000/ha, and those of small-scale schemes were US$2 300–3 400/ha. These costs are prohibitive in most cases and largely explain why the irrigated area in Ethiopia is less than 8 percent of the land considered potentially irrigable (FAO, 1995).

The development of sustainable dryland systems in Ethiopia is also challenging. Seventy percent of the people live in the mountainous areas, where 60 percent of the land has slopes greater than 16 percent (FAO, 1995). Annual soil losses can reach 300 tonnes/ha. Therefore, soil- and water-management practices that will sustain crop production over the long term will also be costly, but considerably less so than the development of irrigated land. The potential for water harvesting is also great because of the sloping lands and high-intensity precipitation events that result in large amounts of runoff.

INDIA

Dryland farming in India began centuries earlier than in North America. However, there are some striking similarities between the two regions with respect to the scientific study of dryland farming. Hegde (1995) reported that, in 1917, Aiyer had listed the important farmer practices and found them quite similar to those that Campbell (1907) had proposed for the Great Plains in the United States of America in the early 1900s. Field bunding, fall ploughing, frequent intercultivations, drill sowing, and growing drought-resistant crops, such as finger millet, grain sorghum and pearl millet, were some of the practices listed.

Scientific study of dryland farming was initiated by the Government of India in 1923. Early research focused on improving crop yields. Important practices included: (i) bunding to conserve soil and water; (ii) deep ploughing once in three years for better intake and storage of water; (iii) use of farmyard manure to supply plant nutrients; (iv) use of a low seeding rate; and (v) intercultivation for weed and evaporation control. These practices gave a 15–20 percent increase over the base yields (Hegde, 1995, Singh, 1995). By the mid-1950s, the emphasis had shifted to soil management. Soil conservation research and training centres were established at eight locations, focusing on contour bunding. However, negative results were often obtained because of water accumulation and runoff problems, particularly on Vertisols. Even where yield increases were observed, they were again not more than 15–20 percent above the base yields.

The importance of shorter-duration crops to match the soil-water availability period was recognized in the 1960s. It was also in the mid-1960s that high-yielding hybrids and cultivars became available that were responsive not only to fertilizers but also to management. An All-India Coordinated Research Project for Dryland Agriculture was established, and the research emphasis shifted to a multidisciplinary approach to tackle the problems. Similar efforts were initiated at the International Crops Research Institute for the Semi-Arid Tropics at Hyderabad in 1972.

Although many of the recommended practices for dryland farming in India are similar to those for North America, there are differences. A highly recommended practice for water conservation in India is the use of dust mulch (Hegde, 1995), similar to that recommended in the North America in the early 1900s, which is commonly considered to have contributed to the Dust Bowl. These contrasting recommendations illustrate the importance of recognizing and addressing the differences between semi-arid regions. Semi-arid conditions in North America are very different from those in India. For example, summer fallow has played an important role in North America because some precipitation occurs throughout the year, but at no time does monthly precipitation exceed PET. Many dryland regions in the Great Plains in the United States of America do not have any month with precipitation that even reaches half of PET. In contrast, most dryland areas in India have more than seven months with essentially zero precipitation. They then have a monsoon season of varying length when the precipitation greatly exceeds the potential evapotranspiration for at least a portion of the growing season. Therefore, fallow for storing soil water is not a viable alternative because much of the water saved during the rainy season would be lost during the prolonged dry period. More importantly, there is usually more than enough precipitation during the monsoon season to fully wet the soil profile.

MEDITERRANEAN REGIONS

A Mediterranean-type climate is characterized by the concentration of rainfall in the winter half-year – from November to April in the northern hemisphere, and from May to October in the southern – with drought in summer. In California (the United States of America) and in Chile in particular, winter rainfall may constitute 80–90 percent of the annual precipitation, but this is less common in the Mediterranean basin itself. A winter precipitation exceeding 65 percent of the annual total has been used to define the climate zone. The worldwide distribution of Mediterranean climates is between latitudes 32 and 40 ° north and south of the equator on the west coasts of the continents (Boyce, Tow and Koocheki, 1991). North of the Mediterranean Sea, they extend into higher latitudes and in western Australia into lower latitudes. Major dryland-farming areas with a Mediterranean-type climate are located in Morocco, Algeria, Tunisia, the Libyan Arab Jamahiriya, Syrian Arab Republic, Jordan, Iraq, Turkey, the Islamic Republic of Iran, Chile, Australia, and parts of California and the

Pacific northwest of the United States of America. Annual precipitation values suitable for agriculture are generally considered to be from 250 to 500 mm. Most areas show a wide fluctuation in: total precipitation from year to year; distribution during the season; time of onset of the rainy season; duration of the rainy season; and intensity of rainfall during precipitation events. Except for Australia, where the topography is moderately flat, countries with Mediterranean-type climates are characterized by rugged mountain ranges. Increasing elevation correlates positively and strongly with increasing precipitation. Dryland agriculture has been practised in some of these areas for perhaps 10 000 years or more (Boyce, Tow and Koocheki, 1991).

The West Asia – North Africa region constitutes a large part of the dryland farming areas of the Mediterranean regions. There are two major types of dryland-farming systems in these areas: the wheat-based systems in wetter areas; and the barley-based systems in drier areas. Although there are some variations owing to elevation, soil depth and soil type, the transition between the two systems is generally considered to be the 300 mm isohyet (Boyce, Tow and Koocheki, 1991). Yields are often low as a consequence of low soil fertility, exacerbated by erosion. Fallow is also common in these areas, but it is practised very differently from in other major dryland cereal-cropping areas. Fallows are generally uncultivated weedy fallows. The fallow fields are grazed by animals, with a primary purpose of increasing soil fertility rather than conserving soil water. The main goal of the barley-based farming systems is animal production. Reliance on animals increases with decreasing amount and reliability of rainfall. Animal production in these low-rainfall areas, as in other dryland regions of the world, is an important survival mechanism in subsistence-type agriculture where the risk of crop failure from inadequate rainfall is high.

Farming systems in these regions continue to evolve depending on a variety of pressures, particularly those related to increasing population, increasing shortage of land, increasing mechanization, changing market forces, and a variety of social and political factors (Boyce, Tow and Koocheki, 1991). In the higher-rainfall wheat-based areas, crop production will probably continue to dominate although there is increasing tree-fruit production. In the lower-rainfall wheat-based systems and the barley-based systems, animal production is assuming greater importance. Barley is replacing both wheat and fallow to satisfy the increasing need for animal-feed production.

Although animal production is increasing, Boyce, Tow and Koocheki (1991) state that an important feature of all of the farming systems is that cropping- and livestock-production systems are not generally integrated. Cropland is predominantly owned and worked by farmers who do not own livestock, while livestock owners do not usually crop the land. This has been the general pattern for centuries or millennia in most of the cropped areas of the Mediterranean basin. This presents both a physical and a social barrier to the use of systems that efficiently integrate cropping and livestock production under one management system, such as the Australian ley farming system.

The cropping systems in the other Mediterranean-type climate regions – Chile, Australia, and parts of California and the Pacific northwest in the United States of America – usually include cereal production as an important component (Boyce, Tow and Koocheki, 1991). However, the mix of crop and farm animals is somewhat different. The degree of industrialization accounts for many of the differences between the farming systems. Broadly speaking, the higher the national income per capita is, so the more intensive and specialized the farming patterns are.

NORTH AMERICA

The Prairies in Canada and the Great Plains, Pacific northwest, parts of the southwest and intermountain areas in the United States of America constitute the major dryland-farming areas of North America. These areas are major contributors to the food and fibre production system. Gras (1946) suggested that, historically, six general types of farming developed in the United States of America. Listed in chronological order, they were: woodland, prairie, ranch, irrigated, dryland, and leftover farming. This is a natural progression that is typical of other regions in the world in that agriculture expands to increasingly marginal lands. In the United States of America, as settlers pushed west from the sub-humid eastern edge of the Great Plains, they found that crop production became more erratic and precarious.

Early-day conservationists warned of the erosion that would take place in many parts of the Great Plains if the grass cover on the land were destroyed by cultivation. However, high wheat prices following the First World War, coupled with the development of power machinery, led to the rapid expansion of cultivated land and large-scale production of wheat and other crops. This expansion in cultivated land took place primarily between 1915 and 1925, but dryland farming had become well established in the early 1900s. One of the early promoters of dryland farming was Campbell (1907), who published a manual of recommended practices that were widely followed. Campbell believed erroneously that a "dust mulch" on the surface not only conserved substantial amounts of soil water but also attracted it from the atmosphere. There was above-average precipitation during the first 30 years of settlement in the Great Plains when these practices were being advocated, and little damage was noticed. Another factor that led to expansion was that the United States Department of Agriculture (USDA) was formed in 1862 and provided a basis for scientific study. This later led to development of a series of dryland experiment stations throughout the Great Plains, some as early as 1903. In 1914, the USDA Division of Dryland Agriculture was established, consisting of 22 field stations to study crop adaptation and cultural practices (map in Burnett, Stewart and Black, 1985). Although a few of these stations remain today as part of the USDA Agricultural Research Service and a few others as units of the Agricultural Experiment Stations of various states, the Division of Dryland Agriculture was terminated in 1938.

After the unusually wet conditions in the Great Plains during the early years of development, a major drought in the 1930s led to severe wind erosion. All of the Great Plains and the Canadian provinces were affected but the worst-hit areas were in southeast Colorado, southwest Kansas, western Oklahoma and northwest Texas, resulting in the infamous Dust Bowl. Surveys made shortly after the drought years of the 1930s in the southern Great Plains wind-erosion area suggested that 43 percent of the area had serious wind erosion damage (Joel, 1937). Finnell (1948a) estimated that about 2.6 million ha in the southern Great Plains were removed from cultivation and remained idle or were returned to grass because erosion had made them unproductive, but that about 10.3 million ha of less erodible soils survived the drought cycle without significant erosion damage and were put back into cultivation in the 1940s. Finnell (1948b) pointed out that 59 percent of the serious erosion had occurred on poorer lands that probably should never have been cultivated. This reinforces the point made by Mathews and Cole (1938) about deciding whether or not particular lands in semi-arid areas should be cultivated.

As already discussed, summer fallow was an important practice in the development of dryland farming in North America. Fallow continues to be an important component of many cropping systems. However, its length has been shortened and the use of crop residues as surface mulches has been found very beneficial in increasing the proportion of precipitation stored in the soil profile for subsequent crops. Conservation tillage and no-tillage systems have proved extremely important for water conservation and protection against wind erosion.

WEST AFRICA

Sub-Sahelian West Africa lies between 10 and 14 °N of the equator and is largely semi-arid, with 300–800 mm annual rainfall. Soils are mostly Yermosols (Aridisols), Cambisols (Inceptisols) and Luvisols (Alfisols) (Lal, 1993). These soils are highly susceptible to wind and water erosion. Estimates of annual soil loss range from 10 to 50 tonnes/ha for wind alone. Much of the dryland farming in West Africa occurs in the sub-Sahelian region. Rains tend to be intense and erratic, droughts are frequent and may persist for several consecutive years. The erratic character of the rains makes dryland cropping very unreliable despite relatively high rainfall during parts of the year. Sorghum and millet are the principal crops in the African summer rainfall zones (Dregne, 1982). Dregne also reported that the dominant soil problems in the sub-Sahelian region are low inherent soil fertility caused by coarse textures, soil acidity, low nutrient holding capacity, and high phosphorus fixation. Other important problems relate to surface sealing as well as compaction and hardening to a depth of several centimetres. Lal (1993) reported that cultivation and accelerated soil erosion leads to a rapid loss of soil organic matter and plant-nutrient reserves, which results in a serious decline in soil productivity.

Several traditional techniques are used to increase water-use efficiency in West Africa. For example, the zai system in Burkina Faso and Mali consists of small pits, about 10 cm deep and 10–30 cm in diameter, at intervals of about 1 m. After mixing in some manure and dead weeds, seeds are planted in the pits at the start of the cropping season. Zai are often combined with contour bunds to capture still more of the rainfall for infiltration into the pits.

Pieri (1995) summarized long-term experiments in semi-arid francophone Africa and concluded that continuous cropping without any fallow period is technically achievable while increasing yields. However, annual application of mineral fertilizer with periodic liming is required. A combination of mineral fertilizer, periodic liming, and periodic application of cattle manure brings several benefits, including higher and more-stable yields. Yield stability is particularly improved where water-harvesting techniques are used in combination with fertilizer applications.

Background tables

TABLE 1
Production of cereals in various regions (1000 tonnes)

	1961	1970	1980	1990	2000	2004
TOTAL CEREALS						
World	877 776	1 193 377	1 550 883	1 952 166	2 061 054	2 221 119
Developed countries	355 141	430 321	599 666	692 307	826 925	829 977
Developing countries	522 635	763 055	951 216	1 259 859	1 234 129	1 391 143
Least Developed countries	48 057	56 520	69 369	82 858	120 970	154 916
North America	180 350	215 421	311 249	369 205	393 846	397 456
European Union (27)	126 953	161 256	226 932	252 204	277 700	269 239
Asia	449 251	648 852	803 313	1 081 779	996 292	1 102 274
Latin America & Caribbean	47 399	71 365	88 443	99 084	137 988	154 677
Africa	46 277	60 472	72 611	93 410	112 608	145 892
Oceania	9 566	13 515	17 199	23 949	35 340	17 176
WHEAT						
World	222 388	310 803	440331	592 372	586 063	605 946
Developed countries	101 202	121 138	195 203	253 853	294 143	286 865
Developing countries	121 186	189 665	245 128	338 518	291 920	319 081
Least Developed countries		2 809	4 393	4 345	6 482	9 429
North America	41 252	45 808	84 092	106 392	87 293	84 575
European Union (27)		58 408	87 041	115 462	132 428	126 249
Asia	108 322	170 547	221 698	305 013	254 528	272 185
Latin America & Caribbean	9 527	11 510	15 091	20 767	23 698	22 636
Africa	5 118	8 081	8 922	13 689	14 382	25 096
Oceania	6 981	8 177	11 162	15 254	22 434	10 075
MAIZE						
World	205 017	265 844	396 685	483 359	593 225	695 228
Developed countries	116 097	145 493	226 526	253 667	322 902	354 150
Developing countries	88 920	120 352	170 159	229 691	270 323	341 078
Least Developed countries	6 630	6 882	9 064	12 008	18 683	23 528
North America	92 130	108 105	174 400	208 598	258 808	276 866
European Union (27)	18 502	29 366	41 346	36 981	51 552	55 782
Asia	48 726	62 404	96 969	142 350	149 063	203 025
Latin America & Caribbean	24 183	38 098	45 058	49 636	76 216	91 778
Africa	16 124	19 880	28 131	37 700	45 031	46 260
Oceania	171	253	312	387	601	558

TABLE 1 (*continued*)
Production of cereals in various regions (1000 tonnes)

	1961	1970	1980	1990	2000	2004
SORGHUM						
World	41 632	56 383	57 870	56 802	55 832	56 485
Developed countries	12 448	18 345	16 313	16 051	14 830	8 704
Developing countries	29 184	38 038	41 557	40 751	41 002	47 782
Least Developed countries	5 083	5 393	6 232	5 613	8 867	14 613
North America	12 198	17 353	14 716	14 562	11 952	7 050
European Union (27)	54	401	633	511	652	570
Asia	16 301	18 793	19 223	18 688	11 148	10 691
Latin America & Caribbean	2 193	7 533	9 322	10 082	11 383	10 973
Africa	10 692	11 710	13 009	11 980	18 467	26 113
Oceania	163	549	925	947	2120	1001
MILLET						
World	25 755	33 333	24 937	30 001	27 747	31 781
Developed countries	285	251	151	240	1 832	1 086
Developing countries	25 471	33 082	24 786	29 761	25 915	30 695
Least Developed countries	3 462	4 421	4 537	5 411	6 722	10 210
North America	112	150	120	180	199	300
European Union (27)	66	48	13	13	26	42
Asia	18 704	24 970	17 301	19 029	13 102	12 891
Latin America & Caribbean	261	125	188	65	48	16
Africa	6 577	7 997	7 298	10 668	12 765	17 788
Oceania	22	34	14	39	57	35

TABLE 2
Irrigated land, area and as percentage of arable land

| World/Continent | Irrigated Land | | | | | |
| | Area (1000 ha) | | | As % of arable land | | |
	1980	1990	2002	1980	1990	2002
WORLD	210 222	244 988	276 719	15.7	17.6	19.7
Developed countries	58 926	66 286	68 060	9.1	10.2	11.1
Industrialized countries	37 355	39 935	43 669	9.9	10.5	11.9
Transition economies	21 571	26 351	24 391	7.9	9.8	10.0
Developing countries	151 296	178 702	208 659	21.9	24.1	26.3
Latin America & the Caribbean	13 811	16 794	18 622	10.8	12.5	12.6
Near East & North Africa	17 982	24 864	28 642	21.8	28.8	32.3
Sub-Saharan Africa	3 980	4 885	5 225	3.2	3.7	3.6
East & Southeast Asia	59 722	65 624	74 748	37.0	33.9	35.1
South Asia	55 798	66 529	81 408	28.6	33.9	41.7
Oceania developing	3	6	14	0.7	1.2	2.4
Continental groupings						
Africa	9 491	11 235	13 400	6.0	6.7	7.0
Asia	132 377	155 009	193 869	31.3	33.8	37.9
Caribbean	1 074	1 269	1 308	22.0	23.3	26.5
Latin America	12 737	15 525	17 314	10.4	12.0	12.1
North America	21 178	21 618	23 285	9.1	9.3	10.5
Oceania	1 686	2 118	2 844	3.6	4.2	5.6
Europe	14 479	17 414	25 220	11.5	14.0	8.8

Source: FAO 2008

TABLE 3
Water-balance values for annual wheat crops at three semi-arid locations

| | Texas (USA) | | | Shaanxi (China) | | | New South Wales (Australia) | | |
	Wheat	Fallow [2]	Total	Wheat	Fallow [2]	Total	Wheat	Fallow [2]	Total
Precipitation (mm)	256	202	458	181	213	394	280	280	560
Evapotranspiration (ET) (mm)	293		293	264		264	360		360
Soil water change [1] (mm)	-37	37		-83	83		-80	80	
Evaporation and runoff (mm)		165	165		130	130		200	200
Potential ET (PET) (mm)	1 140	740	1 880	475	408	883	—	—	—
ET/PET (%)	26			56			—		—
Precipitation/PET (%)		24			45				
ET/Precipitation (%)		64			67			64	
Yield (kg/ha)	900			1 250			2400		
Water-use efficiency (kg/m³)	0.33			0.47			0.67		

[1] The change in plant-available soil water between time of seeding and time of harvest.
[2] Period between time of harvest and seeding of the subsequent wheat crop.

Source: Adapted from Jones, unpublished data; Cornish and Pratley, 1991.

TABLE 4
**Straw-mulch effects on soil-water storage during an 11-month fallow,
water-storage efficiency and dryland grain sorghum yield at Bushland, USA**

Mulch rate	Water storage [1]	Storage efficiency [2]	Grain yield	Total crop water use [3]	WUE [4]
(tonnes/ha)	(mm)	(%)	(tonnes/ha)	(mm)	(kg/m³)
0	72 c [5]	22.6 c	1.78 c	320	0.56
1	99 b	31.1.b	2.41 b	330	0.73
2	100 b	31.4 b	2.60 b	353	0.74
4	116 b	36.5 b	2.98 b	357	0.84
8	139 a	43.7 a	3.68 a	365	1.01
12	147 a	46.2 a	3.99 a	347	1.15

[1] Water use determined to a depth of 1.8 m; precipitation averaged 318 mm in the fallow period.
[2] Storage efficiency is water stored in the soil at the end of the fallow period as a percentage of precipitation occurring during the 11-month fallow period.
[3] Growing-season precipitation plus change in soil water during the growing season.
[4] Water-use efficiency (WUE) based on grain produced, growing-season precipitation and soil-water change during the growing season.
[5] Column values followed by the same letter are not significantly different at the 5-percent level (Duncan's Multiple Range Test).

Source: Adapted from Unger, 1978.

Cereals production maps

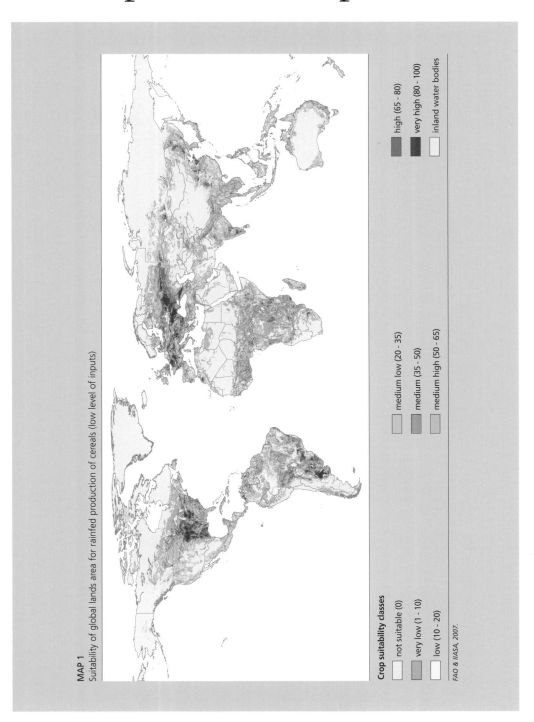

MAP 1
Suitability of global lands area for rainfed production of cereals (low level of inputs)

Crop suitability classes

not suitable (0)	medium low (20 - 35)	high (65 - 80)
very low (1 - 10)	medium (35 - 50)	very high (80 - 100)
low (10 - 20)	medium high (50 - 65)	inland water bodies

FAO & IIASA, 2007.

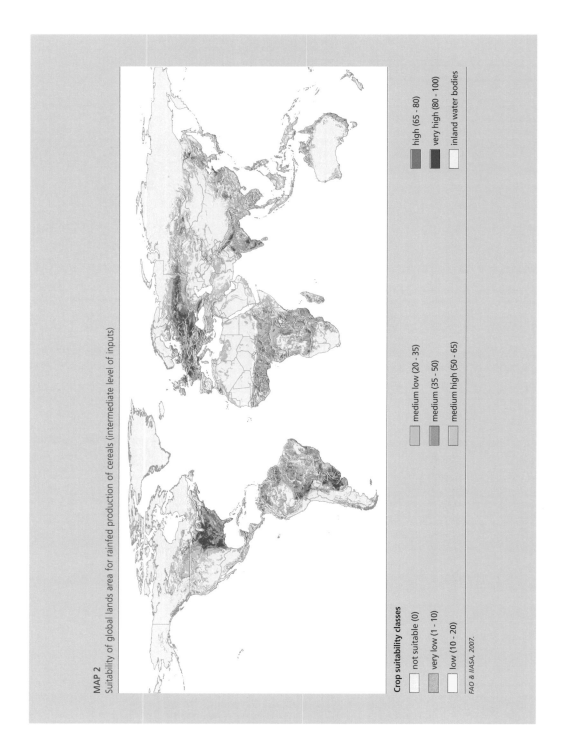

MAP 2
Suitability of global lands area for rainfed production of cereals (intermediate level of inputs)

Crop suitability classes

not suitable (0)

very low (1 - 10)

low (10 - 20)

medium low (20 - 35)

medium (35 - 50)

medium high (50 - 65)

high (65 - 80)

very high (80 - 100)

inland water bodies

FAO & IIASA, 2007.

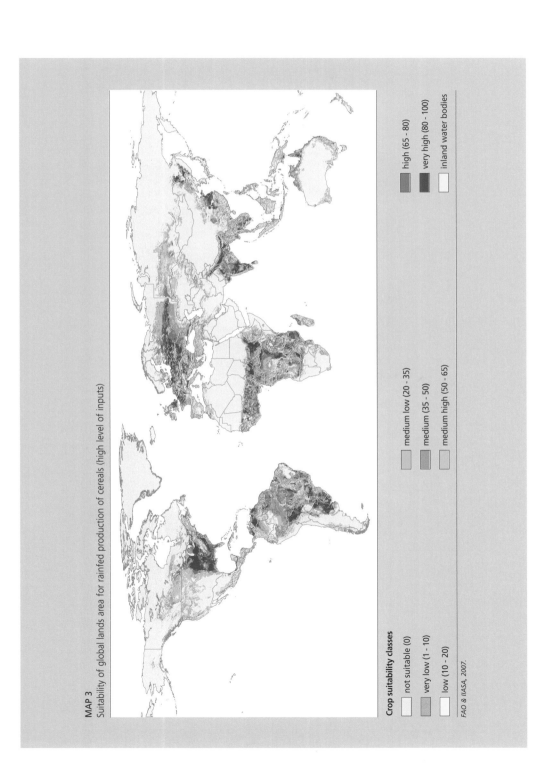

MAP 3
Suitability of global lands area for rainfed production of cereals (high level of inputs)

Crop suitability classes

☐ not suitable (0)

☐ very low (1 - 10)

☐ low (10 - 20)

☐ medium low (20 - 35)

☐ medium (35 - 50)

☐ medium high (50 - 65)

☐ high (65 - 80)

☐ very high (80 - 100)

☐ inland water bodies

FAO & IIASA, 2007.

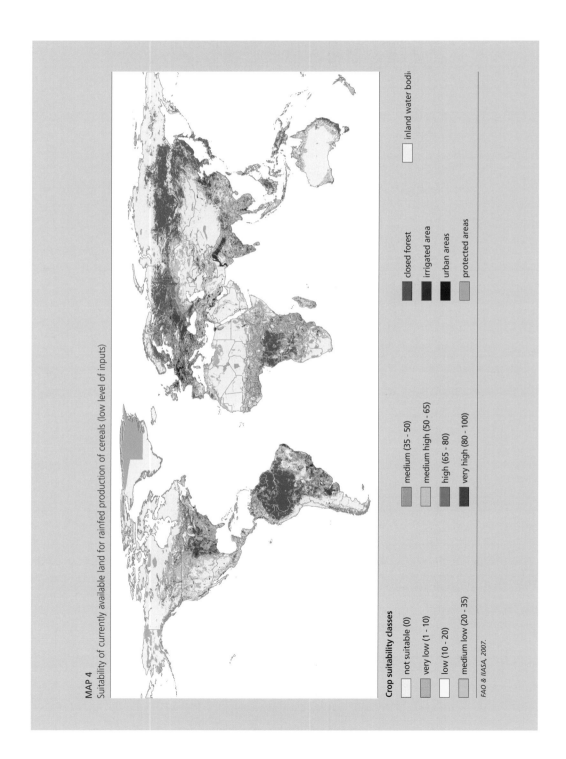

MAP 4
Suitability of currently available land for rainfed production of cereals (low level of inputs)

Crop suitability classes

not suitable (0)

very low (1 - 10)

low (10 - 20)

medium low (20 - 35)

medium (35 - 50)

medium high (50 - 65)

high (65 - 80)

very high (80 - 100)

closed forest

irrigated area

urban areas

protected areas

inland water bodi

FAO & IIASA, 2007.

MAP 5
Suitability of currently available land for rainfed production of cereals (intermediate level of inputs)

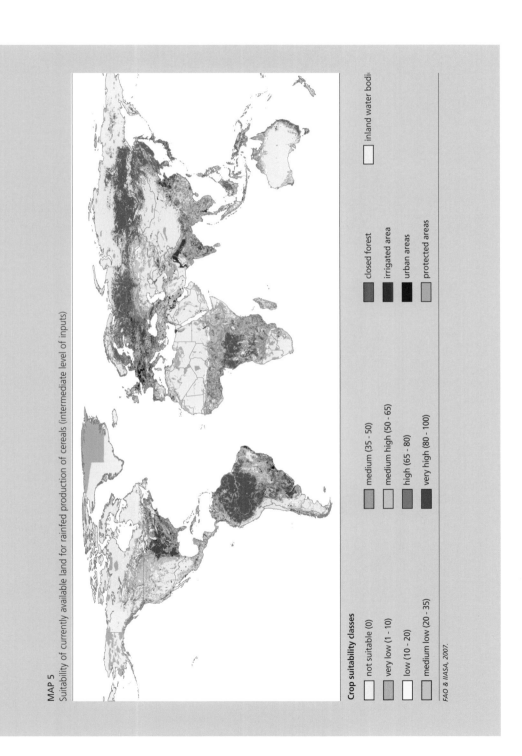

Crop suitability classes

not suitable (0)

very low (1 - 10)

low (10 - 20)

medium low (20 - 35)

medium (35 - 50)

medium high (50 - 65)

high (65 - 80)

very high (80 - 100)

closed forest

irrigated area

urban areas

protected areas

inland water bodi:

FAO & IIASA, 2007.

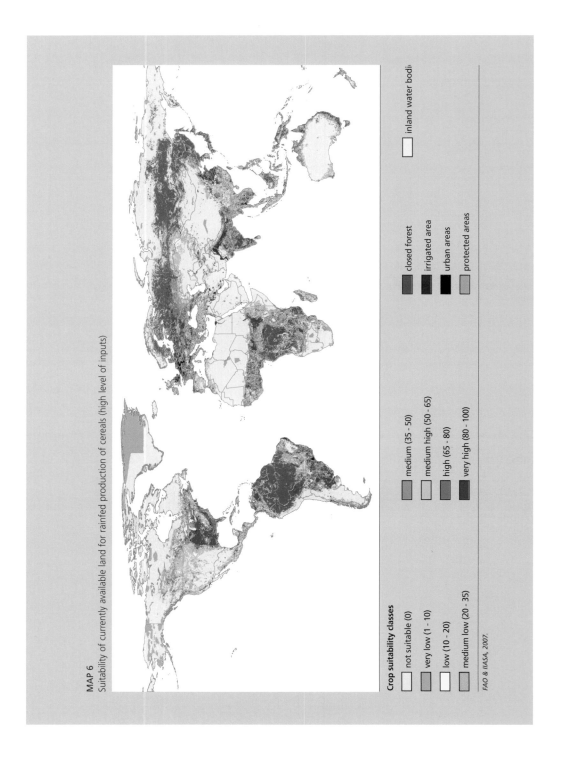

MAP 6
Suitability of currently available land for rainfed production of cereals (high level of inputs)

Crop suitability classes

- not suitable (0)
- very low (1 - 10)
- low (10 - 20)
- medium low (20 - 35)
- medium (35 - 50)
- medium high (50 - 65)
- high (65 - 80)
- very high (80 - 100)

- closed forest
- irrigated area
- urban areas
- protected areas

- inland water bodi

FAO & IIASA, 2007.

MAP 7
Variability of rainfed cereal production potential, by country, 1961–1990

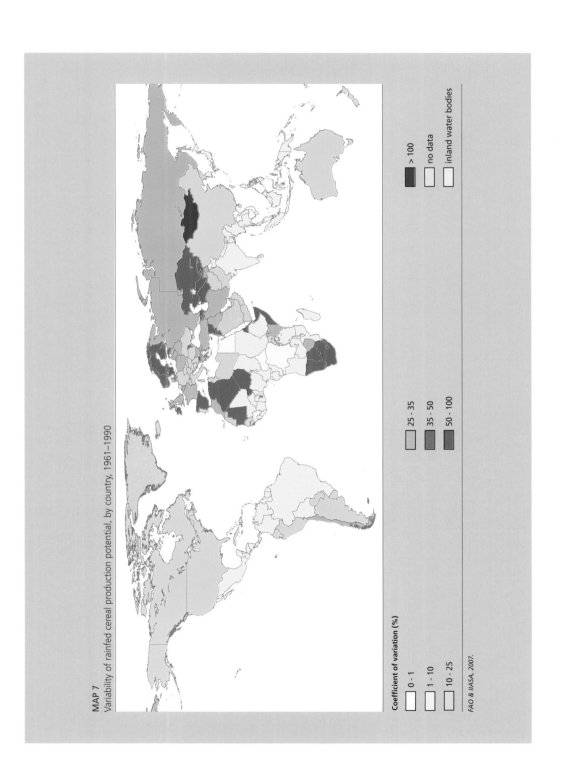

Coefficient of variation (%)

0 - 1
1 - 10
10 - 25
25 - 35
35 - 50
50 - 100
> 100
no data
inland water bodies

FAO & IIASA, 2007.